SpringerBriefs in Mathematics

SpringerBriefs present concise summaries of cutting-edge research and practical applications across a wide spectrum of fields. Featuring compact volumes of 50 to 125 pages, the series covers a range of content from professional to academic. Briefs are characterized by fast, global electronic dissemination, standard publishing contracts, standardized manuscript preparation and formatting guidelines, and expedited production schedules.

Typical topics might include:

- A timely report of state-of-the art techniques
- A bridge between new research results, as published in journal articles, and a contextual literature review
- A snapshot of a hot or emerging topic
- An in-depth case study
- A presentation of core concepts that students must understand in order to make independent contributions

SpringerBriefs in Mathematics showcases expositions in all areas of mathematics and applied mathematics. Manuscripts presenting new results or a single new result in a classical field, new field, or an emerging topic, applications, or bridges between new results and already published works, are encouraged. The series is intended for mathematicians and applied mathematicians. All works are peer-reviewed to meet the highest standards of scientific literature.

Titles from this series are indexed by Scopus, Web of Science, Mathematical Reviews, and zbMATH.

Daniel Goodair • Dan Crisan

Stochastic Calculus in Infinite Dimensions and SPDEs

 Springer

Daniel Goodair
Imperial College London
London, UK

Dan Crisan
Department of Mathematics
Imperial College
London, UK

ISSN 2191-8198   ISSN 2191-8201   (electronic)
SpringerBriefs in Mathematics
ISBN 978-3-031-69585-8   ISBN 978-3-031-69586-5   (eBook)
https://doi.org/10.1007/978-3-031-69586-5

Mathematics Subject Classification: 60H15, 60H05, 60G44, 35R60, 35R15, 35Q35, 60H30

This Springer imprint is published by the registered company Springer Nature Switzerland AG
The registered company address is: Gewerbestrasse 11, 6330 Cham, Switzerland

If disposing of this product, please recycle the paper.

Preface

The purpose of this brief is to cover the basics of infinite dimensional stochastic differential equations (defined on Hilbert spaces), in a pedagogical manner, assuming only an elementary understanding of functional analysis and probability theory. We present a robust construction of the stochastic integral in Hilbert Spaces, considering integrals driven at first by real valued martingales and later by *Cylindrical Brownian Motion*, introducing this concept and expanding into a basic set-up for Stochastic Partial Differential Equations (SPDEs). The framework that we establish facilitates a broad class of SPDEs and noise structures, notably including *unbounded* noise, in which we build upon standard Stochastic Differential Equation (SDE) theory and rigorously deduce a conversion between their Stratonovich and Itô forms. In the remainder of the brief, we explore more advanced tools to be used in the analysis of these equations.

London, UK   Daniel Goodair
July 2024   Dan Crisan

Acknowledgements

Daniel Goodair was supported by the Engineering and Physical Sciences Research Council (EPSCR) Project 2478902. Dan Crisan was partially supported by the European Research Council (ERC) under the European Union's Horizon 2020 Research and Innovation Programme (ERC, Grant Agreement No 856408).

Contents

Chapter 1
Introduction

This chapter introduces the brief, outlining its structure and motivating the core themes. We also establish and collect notation used throughout the book.

1.1 Motivation and Description of the Brief

This brief comprises three chapters, increasing in complexity, described below:

- In Chap. 2 we present a "classical" construction of the Itô stochastic integral, for processes evolving in a Hilbert space. This is introduced first for a one dimensional driving Brownian motion, before generalizations to other one dimensional martingales and, further, to *cylindrical Brownian motion*. Our construction is direct and designed to be familiar to a reader who has undertaken the real valued study as covered, for example, in [44, 55]. In defining the infinite dimensional Brownian motion, that is the cylindrical Brownian motion, we cover the fundamentals of martingale theory in Hilbert spaces broadly by finite dimensional projections along with the real valued theory. The hope is again that this approach is entirely accessible to a reader with background in the real valued integration theory. Precise attention must be paid to the martingale theory in order to properly consider Stratonovich equations; in our opinion the most thorough presentation of this material is in [60], yet more details and results are needed, such as the cross-variation between a Hilbert space valued and real valued martingale. The cylindrical Brownian motion is the only infinite dimensional driving process that we integrate with respect to; while we present a background on general Q-cylindrical processes which could be viable integrators, limiting ourselves to cylindrical Brownian motion enables the integral to be established as a straightforward limit of the integrals against finite dimensional Brownian motions. In particular we avoid the operator theoretic technicalities necessary in the general case, present in the all of the constructions of the classical works

D. Goodair, D. Crisan, *Stochastic Calculus in Infinite Dimensions and SPDEs*, SpringerBriefs in Mathematics, https://doi.org/10.1007/978-3-031-69586-5_1

[20, 38, 51, 52, 57, 60]. We hope that removing some generality makes our approach more accessible for newcomers to the field, and we note that our construction is sufficient for the framework and applications that follow.

- Chapter 3 details a framework for the study of stochastic partial differential equations (SPDEs), which are evolution equations involving integrals of the form introduced in the previous chapter. Through this framework we define notions of solutions for an abstract SPDE, motivated in particular by the recent attention given to *transport type noise* (where the stochastic integral is dependent on the gradient of the solution) and *Stratonovich* equations. Motivation for such study is given below this list of contents. A rigorous mathematical understanding of these equations presents difficulty for two key reasons. The first is the gradient dependency in the noise, taking us beyond the most general "variational frameworks" seen in the literature as these are posed for a noise operator which is bounded on some Hilbert space. The second is the Stratonovich integration, which we are likely to only understand as a corrected Itô integral, yet this conversion is highly nontrivial for a noise which is not bounded on a Hilbert space. Furthermore we wish to consider *nonlinear* SPDEs, such as the evolution equations of fluid dynamics, rendering the well-established linear theory insufficient.

  To be precise, we present a framework which shares its spirit with the variational approach to SPDEs pioneered by Pardoux in the 1970s and now best represented in the more recent books [51, 56, 57]. This classical framework considers an evolution equation with respect to a Gelfand Triple, say $V \hookrightarrow H \hookrightarrow V^*$, where solutions have paths which are square integrable in V, continuous in H, and satisfying an identity in V^*. Recalling our motivation of fluid equations, the prototypical example in this framework is the Navier–Stokes equation. While analytically weak solutions fit this framework seamlessly, analytically strong solutions fit to the spaces $W^{2,2} \hookrightarrow W^{1,2} \hookrightarrow L^2$, which prompts our choice of a triplet of embedded Hilbert spaces without any necessary duality structure. Furthermore, the aforementioned works allow only for a noise operator bounded in H (thus not of first order in applications) and do not consider Stratonovich equations. To include a Stratonovich transport type noise, we introduce a fourth Hilbert space, necessary as the Itô–Stratonovich correction requires an additional derivative to cover the transport type noise. We can then properly define weak, strong, and local solutions of nonlinear PDEs with Stratonovich transport noise, alongside other more classical additive and multiplicative stochastic perturbations. We believe that presenting the technical details, in such generality, of these notions here facilitates the rigorous and free analysis of the equations in future works.

- Chapter 4 contains advanced novel techniques in the existence theory for nonlinear SPDEs. The beauty of the classical variational approach comes from the existence results, which certainly cannot be matched as elegantly in a framework built for 3D Navier–Stokes equations and related stochastic fluid models. Instead we focus on techniques that can be used in this direction, centered around the *Galerkin Method* in which finite dimensional approximations of the SPDE are considered and some properties are used to deduce their limit. Immediately then an existence result for the finite dimensional equations is required, more precisely for where the Hilbert space in which the equation evolves is finite dimensional, but the driving Brownian motion is still infinite dimensional. We assume standard Lipschitz and linear growth conditions, and to the best of our knowledge this result is not present in the literature. There are two predominant ways to deduce the existence of a limit of the finite dimensional approximations, which we detail now.

The first is through *tightness*, which is the stochastic route to relative compactness arguments used in PDE theory. The idea is that from tightness we can deduce relative compactness of the laws of the processes over some suitable function space, at which point Skorohod's Representation Theorem enables the deduction of a limiting process almost surely on a new probability space. Criteria to deduce tightness in relevant function spaces are thus of great significance, and our criteria come largely from the works of [3, 43, 59]. The second is through a Cauchy type argument in the relevant spaces, difficult to execute in the case of local solutions, but recently this has been overcome to great effect due to Glatt-Holtz and Ziane in [33] and extended by the authors here. We defer a greater discussion of this highly technical result to Chap. 4 and emphasize that this is a new result in the cutting edge theory of SPDEs.

An energy equality in this setting is also presented. This is well understood in the typical variational framework, for which we again refer to [51, 57], but we take care in addressing some subtle differences. The first is the loss of the duality structure, though for this result we do assume a bilinear form relation which behaves similarly. The second is that we conduct the proof for *local* solutions, necessary for our motivating class of equations, so it is important for us to explicitly address how the localization affects the proof. Indeed, the consideration of local solutions, as well as the related localization in the construction of the integral, martingale theory, and analytical techniques, is an important extension of the framework of [51, 56, 57]. Similarly, the final key change is that we do not assume any integrability over the probability space of our processes, demanding again another source of localization which we find worthy of detailing in this brief. The chapter rounds out with a demonstration that the infinite dimensional noise can be reduced to one dimensional objects if it is constant multiplicative in each direction.

Before setting up notation and beginning with our exposition, we give some more explicit motivation for this brief. First of all, why work in infinite dimensions? Finite dimensional stochastic differential equations have rich applications in physics and finance, for example, in Langevin equations modeling the movement of a particle in space [13] or the Black–Scholes options pricing model for the dynamics of the price of a stock [8]. These are applications of classical Itô calculus, where the integral of a process takes values in Euclidean spaces. While this theory is adequate in such applications, mathematical models for physical phenomena far exceed those for the position of a particle or that of a tradable stock price. The extension to infinite dimensions is necessary for stochastic differential equations modeling functions of both space and time, such as the velocity or temperature of a fluid. It is therefore necessary to define the stochastic integral

$$\int_0^t \mathbf{\Psi}_s dW_s, \tag{1.1}$$

for a class of stochastic processes $\mathbf{\Psi} : \Omega \times [0, \infty) \times \mathbb{R}^n \rightarrow \mathbb{R}^d$. We regard $\mathbf{\Psi}$ not as a pointwise defined function but rather an element of a function space, which is our motivating context for stochastic integration of Hilbert space valued processes. The recent attention toward Stratonovich and transport noise SPDEs is inspired from the seminal work [42], in which Holm establishes a new class of stochastic equations which serve as fluid dynamics models by adding uncertainty in the transport of fluid parcels to reflect the unresolved scales. For recent literature on the analysis of equations under this stochastic scheme, please see [4, 11, 15–19, 22, 34, 35, 37, 40, 41, 47, 50, 61] to list only a few, all of which rely on an Itô–Stratonovich conversion and a framework such as we present here, which is yet to see any rigorous justification. In fact the pertinence of Stratonovich transport noise in fluid dynamics equations was demonstrated as early as 1992 in the paper [9], and the analysis of SPDEs with general Stratonovich transport noise can be seen in the papers [1, 2, 5, 6, 12, 23–25, 27–31, 39, 48, 49, 53, 54] as well as the recent book [26]. We believe that our framework and results facilitate the rigorous and free analysis of such equations in future works.

1.2   Notation

Throughout the brief, we work with a fixed filtered probability space $(\Omega, \mathcal{F}, (\mathcal{F}_t), \mathbb{P})$, which is complete with respect to $\mathcal{F}_0$. We always consider Banach spaces as measure spaces equipped with the corresponding Borel σ-algebra and shall use λ to denote the Lebesgue Measure. All of our Hilbert Spaces are assumed to be separable.

Notation 1.1 Let (X, μ) denote a general measure space, $(\mathcal{Y}, \|\cdot\|_{\mathcal{Y}})$ and $(\mathcal{Z}, \|\cdot\|_{\mathcal{Z}})$ be Banach Spaces, and $(\mathcal{U}, \langle \cdot, \cdot \rangle_{\mathcal{U}})$, $(\mathcal{H}, \langle \cdot, \cdot \rangle_{\mathcal{H}})$ be general Hilbert spaces:

- $L^p(X; \mathcal{Y})$ is the usual class of measurable p-integrable functions from X into $\mathcal{Y}$, $1 \leq p < \infty$, which is a Banach space with norm

$$\|\phi\|_{L^p(X;\mathcal{Y})}^p := \int_X \|\phi(x)\|_{\mathcal{Y}}^p \mu(dx).$$

The space $L^2(X; \mathcal{Y})$ is a Hilbert Space when $\mathcal{Y}$ itself is Hilbert, with the standard inner product

$$\langle \phi, \psi \rangle_{L^2(X;\mathcal{Y})} = \int_X \langle \phi(x), \psi(x) \rangle_{\mathcal{Y}} \, \mu(dx).$$

- $L^\infty(X; \mathcal{Y})$ *is the usual class of measurable functions from* X *into* $\mathcal{Y}$ *which are essentially bounded, which is a Banach Space when equipped with the norm*

$$\|\phi\|_{L^\infty(X;\mathcal{Y})} := \inf\{C \geq 0 : \|\phi(x)\|_Y \leq C \text{ for } \mu\text{-a.e. } x \in X\}.$$

- $C(X; \mathcal{Y})$ *is the space of continuous functions from* X *into* $\mathcal{Y}$.
- $C_w(X; \mathcal{Y})$ *is the space of "weakly continuous" functions from* X *into* $\mathcal{Y}$, *by which we mean continuous with respect to the given topology on* X *and the weak topology on* $\mathcal{Y}$.
- $\mathscr{L}(\mathcal{Y}; \mathcal{Z})$ *is the space of bounded linear operators from* $\mathcal{Y}$ *to* $\mathcal{Z}$. *This is a Banach Space when equipped with the norm*

$$\|F\|_{\mathscr{L}(\mathcal{Y};\mathcal{Z})} = \sup_{\|y\|_{\mathcal{Y}}=1} \|Fy\|_{\mathcal{Z}}.$$

  $\mathscr{L}(\mathcal{Y}; \mathcal{Z})$ *is the dual space* $\mathcal{Y}^*$ *when* $\mathcal{Z} = \mathbb{R}$, *with operator norm* $\|\cdot\|_{\mathcal{Y}^*}$.
- $\mathscr{L}^1(\mathcal{U}; \mathcal{H})$ *is the space of trace-class operators from* $\mathcal{U}$ *to* $\mathcal{H}$, *defined as the elements* $F \in \mathscr{L}(\mathcal{U}; \mathcal{H})$ *such that for some basis* (e_i) *of* $\mathcal{U}$,

$$\sum_{i=1}^{\infty} \|Fe_i\|_{\mathcal{H}} < \infty.$$

  This is independent of the choice of basis (see, e.g., [14, pp. 267 Ex 20]).
- $\mathscr{L}^2(\mathcal{U}; \mathcal{H})$ *is the space of Hilbert–Schmidt operators from* $\mathcal{U}$ *to* $\mathcal{H}$, *defined as the elements* $F \in \mathscr{L}(\mathcal{U}; \mathcal{H})$ *such that for some basis* (e_i) *of* $\mathcal{U}$,

$$\sum_{i=1}^{\infty} \|Fe_i\|_{\mathcal{H}}^2 < \infty.$$

  This is a Hilbert space with inner product

$$\langle F, G \rangle_{\mathscr{L}^2(\mathcal{U};\mathcal{H})} = \sum_{i=1}^{\infty} \langle Fe_i, Ge_i \rangle_{\mathcal{H}},$$

  which is independent of the choice of basis (see, e.g., [14, pp. 267 Ex 20]).
- *For any* $T > 0$, $\mathscr{S}_T$ *is the subspace of* $C([0, T]; [0, T])$ *of strictly increasing functions.*
- *For any* $T > 0$, $\mathcal{D}([0, T]; \mathcal{Y})$ *is the space of càdlàg functions from* $[0, T]$ *into* $\mathcal{Y}$. *It is a complete separable metric space when equipped with the metric*

$$d(\phi, \psi) := \inf_{\eta \in \mathscr{S}_T} \left[\sup_{t \in [0,T]} |\eta(t) - t| \vee \sup_{t \in [0,T]} \|\phi(t) - \psi(\eta(t))\|_{\mathcal{Y}} \right],$$

which induces the so-called Skorohod topology (see [7, pp 124] for details).
- *The total variation of a function $\phi : [0, T] \to \mathcal{Y}$, $V_{\mathcal{Y}}^T(\phi)$, is defined as*

$$V_{\mathcal{Y}}^T(\phi) := \sup_I \sum_{i=0}^{k-1} \|\phi(t_{i+1}) - \phi(t_i)\|_{\mathcal{Y}}$$

for the supremum taken over all partitions $I = \{0 = t_0 < t_1 < \cdots < t_k = T\}$. A function $\psi : [0, \infty) \to \mathcal{Y}$ is said to be of bounded-variation if $V_{\mathcal{Y}}^T(\psi) < \infty$ for all $T \geq 0$.

For the reader's convenience we also collect notation that is introduced throughout the brief, referencing where it is defined as relevant:

- $\mathbb{1}_A$ denotes the indicator function of the set A.
- $I_T^{\mathcal{H}}$ is defined in Definition 2.4.
- $I^{\mathcal{H}}$ is defined in Definition 2.5.
- $\mathcal{M}^2, \mathcal{M}_c^2$ are defined in Definition 2.7.
- $\mathcal{M}_c^2(\mathcal{H})$ is defined in Definition 2.14.
- $\bar{\mathcal{M}}_c^2, \bar{\mathcal{M}}_c^2(\mathcal{H})$ are the corresponding semimartingale spaces.
- $I_M^{\mathcal{H},T}, I_M^{\mathcal{H}}$ are defined in Definition 2.8.
- $\bar{I}_N^{\mathcal{H},T}, \bar{I}_N^{\mathcal{H}}$ are defined in Definition 2.9.
- $[\cdot]$ is defined in Definition 2.16.
- $[\cdot, \cdot]$ is defined in Definition 2.18.
- $I_T^{\mathcal{H}}(\mathcal{W})$ is defined in Definition 2.19.
- $I^{\mathcal{H}}(\mathcal{W})$ is defined in Definition 2.19.
- $\bar{I}_T^{\mathcal{H}}(\mathcal{W})$ is defined in Definition 2.21.
- $\bar{I}^{\mathcal{H}}(\mathcal{W})$ is defined in Definition 2.21.

Chapter 2
Stochastic Calculus in Infinite Dimensions

In this chapter we present a "classical" construction of the Itô stochastic integral, for processes evolving in a Hilbert space. This is introduced first for a one dimensional driving Brownian motion, before generalizations to other one dimensional martingales and, further, to *cylindrical Brownian motion*. The construction is direct and designed to be familiar to a reader who has undertaken the study of integration with respect to a real valued Brownian motion. In addition, we offer a thorough introduction to martingale theory in Hilbert spaces.

2.1 A Classical Construction for Hilbert Space Valued Processes

The construction of the stochastic integral for Hilbert space valued processes precisely mirrors the standard one dimensional Itô integral; as such we start from simple processes. $\mathcal{H}$ will denote a general separable Hilbert space, with norm and inner product given by $\|\cdot\|_{\mathcal{H}}$ and $\langle \cdot, \cdot \rangle_{\mathcal{H}}$, respectively. W represents a standard one dimensional Brownian motion with respect to the fixed filtered probability space $(\Omega; \mathcal{F}, (\mathcal{F}_t), \mathbb{P})$.

Definition 2.1 Let $0 = t_0 < \cdots < t_i < t_{i+1} < \ldots$ be a partition of the time interval $[0, \infty)$ such that $\lim_{i \to \infty} t_i = \infty$ approaches infinity. A simple $\mathcal{H}$-valued process is one that for $\mathbb{P} - a.e.$ ω takes the form

$$\Psi_t(\omega) = a_0(\omega)\mathbb{1}_{\{0\}}(t) + \sum_{i=0}^{\infty} a_i(\omega)\mathbb{1}_{(t_i, t_{i+1}]}(t),$$

where each $a_i \in L^2(\Omega; \mathcal{H})$ and is $\mathcal{F}_{t_i}$-measurable, with respect to the Borel sigma algebra on $\mathcal{H}$.

© The Author(s), under exclusive license to Springer Nature Switzerland AG 2024
D. Goodair, D. Crisan, *Stochastic Calculus in Infinite Dimensions and SPDEs*,
SpringerBriefs in Mathematics, https://doi.org/10.1007/978-3-031-69586-5_2

Definition 2.2 The Itô integral of a simple $\mathcal{H}$-valued process Ψ, with respect to Brownian motion, is defined as

$$\int_0^t \Psi_s dW_s := \sum_{i=0}^\infty a_i \left(W_{t_{i+1}\wedge t} - W_{t_i \wedge t}\right).$$

Note that in reality the above is a finite sum, so the right hand side is well defined. Indeed, it can alternatively be expressed as

$$\sum_{i=0}^{k-1} a_i \left(W_{t_{i+1}} - W_{t_i}\right) + a_k \left(W_t - W_{t_k}\right), \tag{2.1}$$

where k is such that $t \in (t_k, t_{k+1}]$. We define the integral for a more general class of integrands, using approximations by simple processes. For this we introduce the notion of *progressive measurability*.

Definition 2.3 An $\mathcal{H}$-valued process Ψ is said to be progressively measurable if for every $T > 0$, the restricted process $\Psi : \Omega \times [0, T] \to \mathcal{H}$ is measurable with respect to the product measure $\mathcal{F}_T \times \mathcal{B}([0, T])$.

Definition 2.4 We use $\mathcal{I}_T^{\mathcal{H}}$ to denote the class of $\mathcal{H}$-valued processes Ψ which are progressively measurable and satisfy the square integrability condition

$$\mathbb{E}\left(\int_0^T \|\Psi_s\|_{\mathcal{H}}^2 ds\right) < \infty. \tag{2.2}$$

In other words, $\Psi \in L^2(\Omega \times [0, T]; \mathcal{H})$, where the domain space $\Omega \times [0, T]$ is a measure space equipped with the product measure $\mathbb{P} \times \lambda$.[1]

We have given the definition for progressively measurable processes, not *previsible* processes[2] as will commonly be seen in the literature. Progressive measurability is a weaker condition than previsibility, but thankfully most reasonably behaved processes (adapted and left continuous for example) are both progressively measurable and previsible. We make the definition here for the more general class of integrands *in the cases where the integrator is continuous*. Other authors (e.g., [46]) may opt for previsible processes as these become necessary in retaining nice properties (e.g., martingality) when defining the stochastic integral with respect to discontinuous integrators.

---

[1] The progressive measurability of Ψ ensures that it is measurable over this product space, and Tonelli's theorem justifies exchanging the order of integration.

[2] An $\mathcal{H}$-valued process is said to be previsible if it is measurable with respect to the sigma algebra generated by all left continuous adapted processes.

Definition 2.5 The class of processes $\boldsymbol{\Psi}$ such that $\boldsymbol{\Psi} \in \mathcal{I}_T^{\mathcal{H}}$ for all $T > 0$ is denoted by $\mathcal{I}^{\mathcal{H}}$.

$\mathcal{I}^{\mathcal{H}}$ represents our class of integrands for all times, though there will be nothing wrong with defining the integral for times $t \leq T$ in the class $\mathcal{I}_T^{\mathcal{H}}$. The construction comes as a limit of simple integrals, for which we need the following proposition which holds no differently to [55, Lemma 3.1.5] for example.

Proposition 2.1 *For any $\boldsymbol{\Psi} \in \mathcal{I}_T^{\mathcal{H}}$, there exists a sequence of simple processes $(\boldsymbol{\Psi}^n)$ which converge to $\boldsymbol{\Psi}$ in $L^2(\Omega \times [0, T]; \mathcal{H})$.*

This approximation is the final piece of the constructive jigsaw, allowing us to define the stochastic integral below.

Definition 2.6 We define the Itô stochastic integral for processes $\boldsymbol{\Psi} \in \mathcal{I}^{\mathcal{H}}$ by

$$\int_0^t \boldsymbol{\Psi}_s \, dW_s := \lim_{n \to \infty} \int_0^t \boldsymbol{\Psi}_s^n \, dW_s, \tag{2.3}$$

where $(\boldsymbol{\Psi}^n)$ is the sequence of simple processes postulated in Proposition 2.1 which approach $\boldsymbol{\Psi}$ in $L^2(\Omega \times [0, t]; \mathcal{H})$, and the limit is taken in $L^2(\Omega; \mathcal{H})$.

The fact that this is the natural topology in which to take the limit of simple stochastic integrals falls from the Itô isometry for simple processes, which further justifies that the construction is independent of the choice of simple approximation.

Proposition 2.2 *For a simple process $\boldsymbol{\Psi}^n$ and any time $t > 0$,*

$$\mathbb{E}\left(\left\|\int_0^t \boldsymbol{\Psi}_s^n \, dW_s\right\|_{\mathcal{H}}^2\right) = \mathbb{E}\left(\int_0^t \|\boldsymbol{\Psi}_s^n\|_{\mathcal{H}}^2 \, ds\right). \tag{2.4}$$

Proof Let us suppose that $\boldsymbol{\Psi}^n$ takes the form

$$\boldsymbol{\Psi}_t^n(\omega) = a_0^n(\omega)\mathbb{1}_{\{0\}}(t) + \sum_{i=0}^\infty a_i^n(\omega)\mathbb{1}_{(t_i^n, t_{i+1}^n]}(t) \tag{2.5}$$

as outlined in Definition 2.1. Then applying Definition 2.2, we deconstruct the LHS of (2.4):

$$\mathbb{E}\left(\left\|\int_0^t \boldsymbol{\Psi}_s^n \, dW_s\right\|_{\mathcal{H}}^2\right) = \mathbb{E}\left(\left\|\sum_{i=0}^\infty a_i^n\left(W_{t_{i+1}^n \wedge t} - W_{t_i^n \wedge t}\right)\right\|_{\mathcal{H}}^2\right)$$

$$= \mathbb{E}\left(\left\langle \sum_{i=0}^\infty a_i^n\left(W_{t_{i+1}^n \wedge t} - W_{t_i^n \wedge t}\right), \sum_{j=0}^\infty a_j^n\left(W_{t_{j+1}^n \wedge t} - W_{t_j^n \wedge t}\right)\right\rangle_{\mathcal{H}}\right)$$

$$= \sum_{i=0}^\infty \sum_{j=0}^\infty \mathbb{E}\left(\left\langle a_i^n, a_j^n\right\rangle_{\mathcal{H}} (W_{t_{i+1}^n \wedge t} - W_{t_i^n \wedge t})(W_{t_{j+1}^n \wedge t} - W_{t_j^n \wedge t})\right)$$

recalling once more that the infinite sum is actually a finite sum (2.1) so there is no difficulty in extracting it from the inner product and expectation. For $i \neq j$, and without loss of generality $i < j$, the random inner product is $\mathcal{F}_{t_j}$-measurable as the continuity of the inner product preserves measurability, and therefore $\langle a_i^n, a_j^n \rangle_{\mathcal{H}} (W_{t_{i+1}^n \wedge t} - W_{t_i^n \wedge t})$ and $(W_{t_{j+1}^n \wedge t} - W_{t_j^n \wedge t})$ are independent from the independent increments of Brownian motion. The terms thus vanish, and we are left with

$$\sum_{i=0}^{\infty} \mathbb{E}\left(\|a_i^n\|_{\mathcal{H}}^2 (W_{t_{i+1}^n \wedge t} - W_{t_i^n \wedge t})^2 \right)$$

to which we note independence again and assert that this is just

$$\sum_{i=0}^{\infty} \mathbb{E}(\|a_i^n\|_{\mathcal{H}}^2)(t_{i+1}^n \wedge t - t_i^n \wedge t),$$

which is precisely the integral

$$\int_0^t \sum_{i=0}^{\infty} \mathbb{E}(\|a_i^n\|_{\mathcal{H}}^2) \mathbb{1}_{(t_i^n, t_{i+1}^n]}(s) ds.$$

We can write

$$\|a_i^n\|_{\mathcal{H}}^2 \mathbb{1}_{(t_i^n, t_{i+1}^n]}(s) = \left\| a_i^n \mathbb{1}_{(t_i^n, t_{i+1}^n]}(s) \right\|_{\mathcal{H}}^2,$$

the infinite sum of which is a single nonzero term, equal to

$$\left\| a_0^n(\omega) \mathbb{1}_{\{0\}}(s) + \sum_{i=0}^{\infty} a_i^n(\omega) \mathbb{1}_{(t_i^n, t_{i+1}^n]}(s) \right\|_{\mathcal{H}}^2$$

at every s except for zero which is a set of Lebesgue measure zero in $[0, t]$. Again the infinite sum is truly a single nonzero term, justifying its exchange with expectation, and the above is of course $\|\Psi_s^n\|_{\mathcal{H}}^2$, proving the result.   □

So, why is this useful in terms of the limit in (2.3)? First and foremost, it ensures that the limit is uniquely defined; given that $L^2(\Omega; \mathcal{H})$ is complete, we need only to show that the sequence of stochastic integrals is Cauchy in this space. The Itô isometry tells us that $\left(\int_0^t \Psi_s^n dW_s \right)$ is Cauchy in $L^2(\Omega; \mathcal{H})$ if and only if the sequence (Ψ^n) is Cauchy in $L^2(\Omega \times [0, t]; \mathcal{H})$, which is of course true as by definition the (Ψ^n) are convergent (to Ψ) in this space. Furthermore, the isometry extends to the general integral defined in Definition 2.6, as a trivial corollary of the discussion here.

Corollary 2.1 *The Itô isometry (2.4) holds for all processes* $\boldsymbol{\Psi} \in \mathcal{I}^{\mathcal{H}}$.

Without direct appeal to the formal construction, we may also understand the integral (1.1) as a random element of the dual space $\mathcal{H}^*$ and identify the functional with its counterpart in $\mathcal{H}$ in the usual sense.

Theorem 2.1 *The Itô stochastic integral defined in Definition 2.6 is the unique element of $\mathcal{H}$ satisfying the duality relation*

$$\left\langle \int_0^t \boldsymbol{\Psi}_s dW_s, \phi \right\rangle_{\mathcal{H}} = \int_0^t \langle \boldsymbol{\Psi}_s, \phi \rangle_{\mathcal{H}} dW_s \qquad (2.6)$$

for all $\phi \in \mathcal{H}$. The above are random inner products, defined by

$$\langle \boldsymbol{\Psi}_s, \phi \rangle_{\mathcal{H}}(\omega) := \langle \boldsymbol{\Psi}_s(\omega), \phi \rangle_{\mathcal{H}}$$

and similarly for the LHS.

Remark 2.1 As a corollary, by the Riesz Representation Theorem, it is consistent to define the Itô stochastic integral as an $\mathcal{H}^*$-valued random variable via the mapping

$$\phi \mapsto \left(\int_0^t \langle \boldsymbol{\Psi}_s, \phi \rangle_{\mathcal{H}} dW_s \right)(\omega).$$

Proof Given that we have defined (1.1) as a limit of simple processes, it will come as no surprise that we use this approach to prove the relation (2.6). We will prove that the result holds for simple processes $\boldsymbol{\Psi}^n$ and later that it is preserved in the $L^2(\Omega \times [0, t]; \mathcal{H})$ limit. First though we ought to verify that the RHS of (2.6) makes sense, that is to say $\langle \boldsymbol{\Psi}, \phi \rangle_{\mathcal{H}}$ is a valid (one dimensional) integrand. Thus we must show the standard progressive measurability and square integrability conditions: For the former, note that the progressive measurability of $\boldsymbol{\Psi}$ is preserved under composition with the continuous mapping $\langle \cdot, \phi \rangle_{\mathcal{H}}$. The latter is straightforward, as

$$\mathbb{E}\left(\int_0^T \langle \boldsymbol{\Psi}_s, \phi \rangle_{\mathcal{H}}^2 ds \right) \leq \mathbb{E}\left(\int_0^T \|\boldsymbol{\Psi}_s\|_{\mathcal{H}}^2 \|\phi\|_{\mathcal{H}}^2 ds \right) = \|\phi\|_{\mathcal{H}}^2 \mathbb{E}\left(\int_0^T \|\boldsymbol{\Psi}_s\|_{\mathcal{H}}^2 ds \right),$$
$$(2.7)$$

which is finite by (2.2). Let us suppose that $\boldsymbol{\Psi}^n$ takes the form (2.5). Then applying Definition 2.2, we deconstruct the LHS of (2.6):

$$\left\langle \int_0^t \boldsymbol{\Psi}_s^n dW_s, \phi \right\rangle_{\mathcal{H}} = \left\langle \sum_{i=0}^{\infty} a_i^n \left(W_{t_{i+1}^n \wedge t} - W_{t_i^n \wedge t} \right), \phi \right\rangle_{\mathcal{H}}$$

$$= \sum_{i=0}^{\infty} \left\langle a_i^n \left(W_{t_{i+1}^n \wedge t} - W_{t_i^n \wedge t} \right), \phi \right\rangle_{\mathcal{H}}$$

$$= \sum_{i=0}^{\infty} \langle a_i^n, \phi \rangle_{\mathcal{H}} \left(W_{t_{i+1}^n \wedge t} - W_{t_i^n \wedge t} \right)$$

and proceed similarly for the RHS, observing that the integrand

$$\langle \Psi_s, \phi \rangle_{\mathcal{H}} = \left\langle a_0^n \left(\mathbb{1}_{\{0\}}(s) \right) + \sum_{i=0}^{\infty} a_i^n \left(\mathbb{1}_{(t_i^n, t_{i+1}^n]}(s) \right), \phi \right\rangle_{\mathcal{H}}$$

$$= \left\langle a_0^n \left(\mathbb{1}_{\{0\}}(s) \right), \phi \right\rangle_{\mathcal{H}} + \sum_{i=0}^{\infty} \left\langle a_i^n \left(\mathbb{1}_{(t_i^n, t_{i+1}^n]}(s) \right), \phi \right\rangle_{\mathcal{H}}$$

$$= \langle a_0^n, \phi \rangle_{\mathcal{H}} \left(\mathbb{1}_{\{0\}}(s) \right) + \sum_{i=0}^{\infty} \langle a_i^n, \phi \rangle_{\mathcal{H}} \left(\mathbb{1}_{(t_i^n, t_{i+1}^n]}(s) \right)$$

is again simple. This is completely analogous to showing that $\langle \Psi, \phi \rangle_{\mathcal{H}}$ was a valid integrand. Applying Definition 2.2 to the above proves the result for simple processes, so all that remains to show is preservation in the limit. By definition

$$\left\langle \int_0^t \Psi_s \, dW_s, \phi \right\rangle_{\mathcal{H}} = \left\langle \lim_{n \to \infty} \int_0^t \Psi_s^n \, dW_s, \phi \right\rangle_{\mathcal{H}}$$

and a reminder that this limit is taken in $L^2(\Omega; \mathcal{H})$. We would like to take the limit outside of the inner product, in some appropriate topology, and use the result for simple functions: The steps would be

$$\left\langle \lim_{n \to \infty} \int_0^t \Psi_s^n \, dW_s, \phi \right\rangle_{\mathcal{H}} = \lim_{n \to \infty} \left\langle \int_0^t \Psi_s^n \, dW_s, \phi \right\rangle_{\mathcal{H}}$$

$$= \lim_{n \to \infty} \int_0^t \langle \Psi_s^n, \phi \rangle_{\mathcal{H}} \, dW_s$$

so it should be clear that the topology we want to take this limit in is that of $L^2(\Omega; \mathbb{R})$, as the last line would be precisely the RHS of (2.6) by definition if we can show that the simple real valued process $\langle \Psi^n, \phi \rangle_{\mathcal{H}}$ converges to $\langle \Psi, \phi \rangle_{\mathcal{H}}$ in $L^2(\Omega \times [0, t]; \mathbb{R})$. Thankfully, it is straightforward to justify taking this limit outside of the inner product: If (f_n) converges to f in $L^2(\Omega; \mathcal{H})$, then

$$\mathbb{E}\left(\langle f_n, \phi \rangle_{\mathcal{H}} - \langle f, \phi \rangle_{\mathcal{H}} \right)^2 = \mathbb{E}\left(\langle f_n - f, \phi \rangle_{\mathcal{H}} \right)^2 \le \mathbb{E}\left(\|f_n - f\|_{\mathcal{H}}^2 \|\phi\|_{\mathcal{H}}^2 \right)$$

$$= \|\phi\|_{\mathcal{H}}^2 \mathbb{E}\left(\|f_n - f\|_{\mathcal{H}}^2 \right) \longrightarrow 0$$

so $(\langle f_n, \phi \rangle_{\mathcal{H}})$ converges to $\langle f, \phi \rangle_{\mathcal{H}}$ in $L^2(\Omega; \mathbb{R})$, as required to justify the interchange. To show the convergence of $\langle \Psi^n, \phi \rangle_{\mathcal{H}}$ to $\langle \Psi, \phi \rangle_{\mathcal{H}}$ in $L^2(\Omega \times [0, t]; \mathbb{R})$, we apply the same argument:

$$\left\| \langle \Psi, \phi \rangle_{\mathcal{H}} - \langle \Psi^n, \phi \rangle_{\mathcal{H}} \right\|_{L^2(\Omega \times [0,t]; \mathbb{R})} = \left\| \langle \Psi - \Psi^n, \phi \rangle_{\mathcal{H}} \right\|_{L^2(\Omega \times [0,t]; \mathbb{R})}$$

$$= \mathbb{E}\left(\int_0^t \langle \Psi_s - \Psi_s^n, \phi \rangle_{\mathcal{H}}^2 ds \right)$$

$$\leq \mathbb{E}\left(\int_0^t \| \Psi_s - \Psi_s^n \|_{\mathcal{H}}^2 \| \phi \|_{\mathcal{H}}^2 ds \right)$$

$$= \| \phi \|_{\mathcal{H}}^2 \mathbb{E}\left(\int_0^t \| \Psi_s - \Psi_s^n \|_{\mathcal{H}}^2 ds \right)$$

$$= \| \phi \|_{\mathcal{H}}^2 \left\| \Psi - \Psi^n \right\|_{L^2(\Omega \times [0,t]; \mathcal{H})}$$

$$\longrightarrow 0,$$

where convergence to 0 is by definition of the approximating sequence Ψ^n. □

We provide two applications of this result below, both of which will be fundamental to our SPDE framework.

Proposition 2.3 *The Itô isometry holds for a multidimensional driving Brownian motion, in the sense that if $(\Psi^i)_{i=1}^n$ are a collection of processes in $\mathcal{I}^{\mathcal{H}}$, and $(W^i)_{i=1}^n$ are independent Brownian motions, then*

$$\mathbb{E}\left(\left\| \sum_{i=1}^n \int_0^t \Psi_s^i dW_s^i \right\|_{\mathcal{H}}^2 \right) = \sum_{i=1}^n \mathbb{E}\left(\int_0^t \| \Psi_s^i \|_{\mathcal{H}}^2 ds \right).$$

Proof We look to simplify the left hand side of the required equality, applying Parseval's identity (e.g., [45, pp. 170]) for a basis (e_k) of $\mathcal{H}$:

$$\mathbb{E}\left(\left\| \sum_{i=1}^n \int_0^t \Psi_s^i dW_s^i \right\|_{\mathcal{H}}^2 \right) = \mathbb{E} \sum_{k=1}^\infty \left\langle \sum_{i=1}^n \int_0^t \Psi_s^i dW_s^i, e_k \right\rangle_{\mathcal{H}}^2$$

$$= \mathbb{E} \sum_{k=1}^\infty \left(\sum_{i=1}^n \int_0^t \langle \Psi_s^i, e_k \rangle_{\mathcal{H}} dW_s^i \right)^2$$

having used linearity of the inner product to pull out the sum and Theorem 2.1. We can now regard the infinite sum as an integral with respect to the counting measure and apply Tonelli's theorem, also expanding the square to obtain

$$\sum_{k=1}^\infty \mathbb{E}\left(\sum_{i=1}^n \int_0^t \langle \Psi_s^i, e_k \rangle_{\mathcal{H}} dW_s^i \right)^2$$

$$= \sum_{k=1}^\infty \mathbb{E} \sum_{i=1}^n \sum_{j=1}^n \left(\int_0^t \langle \Psi_s^i, e_k \rangle_{\mathcal{H}} dW_s^i \right)\left(\int_0^t \langle \Psi_s^j, e_k \rangle_{\mathcal{H}} dW_s^j \right).$$

For the cross terms $i \neq j$ we make use of the independence of the Brownian motions and hence the respective stochastic integrals, as well as the standard property that the Itô integral has zero expectation to nullify these terms. Our expression reduces to

$$\sum_{k=1}^{\infty} \sum_{i=1}^{n} \mathbb{E} \left(\int_0^t \left\langle \boldsymbol{\Psi}_s^i, e_k \right\rangle_{\mathcal{H}} dW_s^i \right)^2$$

to which we can apply the Itô isometry (Corollary 2.1 for the Hilbert space $\mathbb{R}$, which is of course the standard isometry) giving us

$$\sum_{k=1}^{\infty} \sum_{i=1}^{n} \mathbb{E} \int_0^t \left\langle \boldsymbol{\Psi}_s^i, e_k \right\rangle_{\mathcal{H}}^2 ds.$$

From this, we apply Tonelli twice more to take the infinite sum all the way through,

$$\sum_{i=1}^{n} \mathbb{E} \int_0^t \sum_{k=1}^{\infty} \left\langle \boldsymbol{\Psi}_s^i, e_k \right\rangle_{\mathcal{H}}^2 ds.$$

A final application of Parseval's identity gives the result. $\square$

While we chose to prove Proposition 2.2 and subsequently Corollary 2.1 from first principles in the Hilbert space setting, the method of proof here touches upon a fundamental aspect of this theory: With a good understanding of the standard setting in $\mathbb{R}$, we can apply Theorem 2.1 to straightforwardly deduce key properties here. Indeed, if we accepted the Itô isometry in $\mathbb{R}$, we could have just proven Corollary 2.1 in the simple vein of Proposition 2.3. We take this approach in extending the result of Theorem 2.1.

Theorem 2.2 *Suppose that $\mathcal{H}_1, \mathcal{H}_2$ are Hilbert spaces such that $\boldsymbol{\Psi} \in \mathcal{I}^{\mathcal{H}_1}$ and $T \in \mathscr{L}(\mathcal{H}_1; \mathcal{H}_2)$. Then the process $T\boldsymbol{\Psi}$ defined by*

$$(T\boldsymbol{\Psi})_s(\omega) = T\left(\boldsymbol{\Psi}_s(\omega)\right)$$

belongs to $\mathcal{I}^{\mathcal{H}_2}$ and is such that

$$T\left(\int_0^t \boldsymbol{\Psi}_s dW_s \right) = \int_0^t T\boldsymbol{\Psi}_s dW_s. \tag{2.8}$$

Proof We shall prove first that $T\boldsymbol{\Psi} \in \mathcal{I}^{\mathcal{H}_2}$. The progressive measurability is preserved under the continuity of T, and for C the (square of the) boundedness constant associated to T, we have that at any time t,

$$\mathbb{E}\left(\int_0^t \|T\Psi_s\|_{\mathcal{H}_2}^2 ds\right) \leq C\mathbb{E}\left(\int_0^t \|\Psi_s\|_{\mathcal{H}_1}^2 ds\right) < \infty$$

as $\Psi \in I^{\mathcal{H}_1}$, showing that $T\Psi \in I^{\mathcal{H}_2}$. To commute T with the integral, we shall use the characterization from Theorem 2.1, having now established that the right hand side of (2.8) is well defined in $\mathcal{H}_2$. We introduce $T^* \in \mathcal{L}(\mathcal{H}_2; \mathcal{H}_1)$ as the adjoint of T and observe that for any $\phi \in \mathcal{H}_2$,

$$\left\langle T\left(\int_0^t \Psi_s dW_s\right), \phi \right\rangle_{\mathcal{H}_2} = \left\langle \int_0^t \Psi_s dW_s, T^*\phi \right\rangle_{\mathcal{H}_1}$$

$$= \int_0^t \langle \Psi_s, T^*\phi \rangle_{\mathcal{H}_1} dW_s$$

$$= \int_0^t \langle T\Psi_s, \phi \rangle_{\mathcal{H}_2} dW_s$$

$$= \left\langle \int_0^t T\Psi_s dW_s, \phi \right\rangle_{\mathcal{H}_2}$$

applying Theorem 2.1 twice. As this equality holds for arbitrary $\phi \in \mathcal{H}_2$, then we have proven (2.8), which is of course an identity $\mathbb{P} - a.e.$ in $\mathcal{H}_2$.   □

2.2 Martingale and Local Martingale Integrators

As expected, we can extend the definition to integrators beyond Brownian motion, in the same manner as the standard Itô integral. We begin the extension to continuous square integrable martingales and then to continuous local martingales.

Definition 2.7 We shall denote the class of real valued martingales M such that $M_t \in L^2(\Omega; \mathbb{R})$ for every $t \geq 0$ by $\mathcal{M}^2$. The subclass of such martingales with $\mathbb{P} - a.s.$ continuous paths will be represented by $\mathcal{M}_c^2$.

Definition 2.8 For any $M \in \mathcal{M}_c^2$, define $I_M^{\mathcal{H}, T}$ to be the class of $\mathcal{H}$-valued processes Ψ which are progressively measurable on $[0, T] \times \Omega$ and satisfy the square integrability condition

$$\mathbb{E}\left(\int_0^T \|\Psi_s\|_{\mathcal{H}}^2 d[M]_s\right) < \infty,$$

where $[M]_.$ is the quadratic variation of $M_.$. We similarly denote by $I_M^{\mathcal{H}}$ the class of processes Ψ such that $\Psi \in I_M^{\mathcal{H}, T}$ for all $T > 0$.

Constructing the integral

$$\int_0^t \Psi_s dM_s$$

for $\Psi \in \mathcal{I}_M^{\mathcal{H}}$ now falls from what we have already done for (1.1). We use simple processes Ψ^n as in Definition 2.1 to approximate Ψ, in the sense that

$$\lim_{n \to 0} \mathbb{E}\left(\int_0^T \|\Psi_s - \Psi_s^n\|_{\mathcal{H}}^2 d[M]_s \right) = 0.$$

Simply replacing W by M in Definitions 2.2 and 2.6 completes the construction, so we do not give the details here. Let us now move on to the more delicate matter of integration with respect to a local martingale. This begins again with notation for our set of integrands.

Definition 2.9 For a continuous local martingale N, define $\bar{\mathcal{I}}_N^{\mathcal{H},T}$ to be the class of progressively measurable processes Ψ such that

$$\int_0^T \|\Psi_s\|_{\mathcal{H}}^2 d[N]_s < \infty \quad \mathbb{P} - a.e. \tag{2.9}$$

Also define $\bar{\mathcal{I}}_N^{\mathcal{H}}$ to be those processes Ψ in $\bar{\mathcal{I}}_N^{\mathcal{H},T}$ for every T.

Suppose that N is localized by the stopping times (T_n). Without loss of generality, this sequence of stopping times can be chosen such that the stopped processes N^{T_n} defined by

$$N_t^{T_n} := N_{t \wedge T_n}$$

are bounded; if (T_n') are localizing stopping times, then we can simply set

$$T_n = T_n' \wedge \inf\{0 \le t < \infty : |N_t| \ge n\}$$

so that for each n, N^{T_n} is a bounded continuous martingale and hence in $\mathcal{M}_c^2$. Note of course that the new stopping times (T_n) are still nondecreasing and approach infinity $\mathbb{P} - a.s.$ by the pathwise continuity of N. Continuing in this manner, for a process $\Psi \in \bar{\mathcal{I}}_N^{\mathcal{H}}$, let us define some more nondecreasing random times (R_n) by

$$R_n := n \wedge \inf\{0 \le t < \infty : \int_0^t \|\Psi_s\|_{\mathcal{H}}^2 d[N]_s \ge n\} \tag{2.10}$$

taking the convention that the infimum of the empty set is infinite. The random variables (R_n) are stopping times as they are simply first hitting times of the continuous and adapted processes

$$\int_0^t \|\mathbf{\Psi}_s\|_{\mathcal{H}}^2 d[N]_s.$$

Again these times tend to infinity $\mathbb{P} - a.s.$ by condition (2.9). Now define τ_n by

$$\tau_n = R_n \wedge T_n \qquad (2.11)$$

and the truncated processes $\mathbf{\Psi}^n$ as

$$\mathbf{\Psi}_t^n := \mathbf{\Psi}_t \mathbb{1}_{t \le \tau_n}.$$

Understanding the fact that for $m \le n$, and $t \le \tau_m$, we have $\mathbf{\Psi}_t \mathbb{1}_{t \le \tau_n} = \mathbf{\Psi}_t \mathbb{1}_{t \le \tau_m}$ and also that $N_t^{\tau_n} = N_t^{\tau_m}$, and then eventually the sequence

$$\left(\int_0^t \mathbf{\Psi}_s^n dN_s^{\tau_n} \right)$$

is constant $\mathbb{P} - a.s..$ Thus, we can make the following definition.

Definition 2.10 In the setting described, we define

$$\int_0^t \mathbf{\Psi}_s dN_s := \lim_{n \to \infty} \int_0^t \mathbf{\Psi}_s^n dN_s^{\tau_n} \qquad (2.12)$$

for the limit taken $\mathbb{P} - a.s.$ in $\mathcal{H}$.

Of course to do this we require that at any n, $\mathbf{\Psi}^n \in I_{N^{\tau_n}}^{\mathcal{H}}$: The process $(\mathbb{1}_{t \le \tau_n})$ is progressively measurable, as it is both left continuous and adapted (adaptedness becomes clear when for each fixed t, we write the random variable $\mathbb{1}_{t \le \tau_n}$ as $1 - \mathbb{1}_{t > \tau_n}$). The square integrability in Definition 2.8 comes from the fact that the random variable

$$\int_0^t \|\mathbf{\Psi}_s^n\|_{\mathcal{H}}^2 d[N^{\tau_n}]_s$$

is bounded by n $\mathbb{P} - a.s.$ (owing to (2.10)), and hence, the expectation satisfies the same bound. Where N is itself a genuine martingale, this procedure defines the stochastic integral for processes with only the regularity (2.9). In this case we do not have to stop the integrator, and we just truncate the integrand.

Definition 2.11 In the special case where the continuous local martingale is a Brownian motion, we denote $\bar{I}_W^{\mathcal{H}}$ by simply $\bar{I}^{\mathcal{H}}$. This class of processes differs from $I^{\mathcal{H}}$ due to the assumption $\mathbb{P} - a.s.$, i.e., (2.9), as opposed to (2.2) in expectation.

We extend properties of the stochastic integral to this class of processes.

Proposition 2.4 *Let* $\Psi \in \bar{I}^{\mathcal{H}}$ *and* $\phi \in L^{\infty}(\Omega; \mathcal{H})$ *be* $\mathcal{F}_0$-*measurable. Then* $\langle \Psi, \phi \rangle \in \bar{I}^{\mathcal{H}}$, *and for every* $t > 0$, *we have that*

$$\left\langle \int_0^t \Psi_r dW_r, \phi \right\rangle_{\mathcal{H}} = \int_0^t \langle \Psi_r, \phi \rangle_{\mathcal{H}} dW_r \tag{2.13}$$

$\mathbb{P} - a.s.$. *The above are random inner products defined by*

$$\langle \Psi_s, \phi \rangle_{\mathcal{H}}(\omega) := \langle \Psi_s(\omega), \phi(\omega) \rangle_{\mathcal{H}}$$

and similarly for the left hand side.

Proof We should first justify that $\langle \Psi., \phi \rangle_{\mathcal{H}} \in \bar{I}^{\mathbb{R}}$. The progressive measurability follows as for every $T > 0$ the mapping

$$\langle \Psi., \phi \rangle_{\mathcal{H}} : t \times \omega \times \tilde{\omega} \mapsto \langle \Psi_t(\omega), \phi(\tilde{\omega}) \rangle_{\mathcal{H}}$$

is $\mathcal{B}([0, T]) \times \mathcal{F}_T \times \mathcal{F}_0$ measurable, so in particular it is $\mathcal{B}([0, T]) \times \mathcal{F}_T \times \mathcal{F}_T$ measurable, and as such

$$\langle \Psi., \phi \rangle_{\mathcal{H}} : t \times \omega \mapsto \langle \Psi_t(\omega), \phi(\omega) \rangle_{\mathcal{H}}$$

is $\mathcal{B}([0, T]) \times \mathcal{F}_T$ measurable as required. Note that we have used the progressive measurability requirement on Ψ. We also appreciate that for $\mathbb{P} - a.e.$ ω,

$$\int_0^T \langle \Psi_r(\omega), \phi(\omega) \rangle_{\mathcal{H}}^2 dr \le \|\phi(\omega)\|_{\mathcal{H}}^2 \int_0^T \|\Psi_r(\omega)\|_{\mathcal{H}}^2 dr < \infty$$

again by assumption on $\Psi \in \bar{I}^{\mathcal{H}}$. Thus $\langle \Psi, \phi \rangle_{\mathcal{H}}$ belongs to $\bar{I}^{\mathbb{R}}$. To compute the integrals we introduce the stopping times

$$\tau_j := j \wedge \inf \left\{ 0 \le t < \infty : (1 + \|\phi\|_{\mathcal{H}}^2) \int_0^t \|\Psi_r\|_{\mathcal{H}}^2 dr \ge j \right\}$$

such that for every $j \in \mathbb{N}$, $\Psi.\mathbb{1}_{.\le \tau_j} \in I^{\mathcal{H}}$, $\langle \Psi.\mathbb{1}_{.\le \tau_j}, \phi \rangle_{\mathcal{H}} \in I^{\mathbb{R}}$. It is sufficient to show that

$$\left\langle \int_0^t \Psi_r \mathbb{1}_{r \le \tau_j} dW_r, \phi \right\rangle_{\mathcal{H}} = \int_0^t \langle \Psi_r \mathbb{1}_{r \le \tau_j}, \phi \rangle_{\mathcal{H}} dW_r \tag{2.14}$$

holds $\mathbb{P} - a.e.$ for every j. We now fix an arbitrary $j \in \mathbb{N}$. The plan is as follows: We consider a sequence of simple processes (Φ^n) which approximate $\Psi.\mathbb{1}_{.\le \tau_j}$ in $L^2(\Omega \times [0, t]; \mathcal{H})$ as postulated in Proposition 2.1. We then claim that $(\langle \Phi^n, \phi \rangle_{\mathcal{H}})$ is a sequence of $\mathbb{R}$ valued simple processes which converge to $\langle \Psi.\mathbb{1}_{.\le \tau_j}, \phi \rangle_{\mathcal{H}}$ in

$L^2(\Omega \times [0, t]; \mathbb{R})$. Following this, we prove (2.13) for this simple case and show the identity holds in the limit.

We first show that for each $n \in \mathbb{N}$, $\langle \Phi^n, \phi \rangle_{\mathcal{H}}$ is a simple process. Let Φ^n have representation as in Definition 2.1. Then

$$\langle \Phi^n, \phi \rangle_{\mathcal{H}} = \left\langle a_0^n \mathbb{1}_{\{0\}} + \sum_{i=0}^{\infty} a_i^n \mathbb{1}_{(t_i^n, t_{i+1}^n]}, \phi \right\rangle_{\mathcal{H}}$$

$$= \langle a_0^n, \phi \rangle_{\mathcal{H}} \mathbb{1}_{\{0\}} + \sum_{i=0}^{\infty} \langle a_i^n, \phi \rangle_{\mathcal{H}} \mathbb{1}_{(t_i^n, t_{i+1}^n]}$$

so this would satisfy the requirements of an $\mathbb{R}$ valued simple process if for each $i \in \mathbb{N}$, $\langle a_i^n, \phi \rangle_{\mathcal{H}} \in L^2(\Omega; \mathbb{R})$ and is $\mathcal{F}_{t_i}$-measurable. For the square integrability constraint, observe that

$$\mathbb{E}\left(\langle a_i^n, \phi \rangle_{\mathcal{H}}^2\right) \leq \mathbb{E}\left(\|a_i^n\|_{\mathcal{H}}^2 \|\phi\|_{\mathcal{H}}^2\right) \leq \|\phi\|_{L^{\infty}(\Omega; \mathcal{H})}^2 \mathbb{E}\left(\|a_i^n\|_{\mathcal{H}}^2\right) < \infty$$

by the assumptions of $a_i \in L^2(\Omega; \mathcal{H})$ and $\phi \in L^{\infty}(\Omega; \mathcal{H})$. The $\mathcal{F}_{t_i}$ measurability follows in the same way as the progressive measurability of $\langle \Psi, \phi \rangle_{\mathcal{H}}$. Indeed the required $L^2(\Omega \times [0, t]; \mathbb{R})$ convergence follows similarly as

$$\left\| \langle \Psi.\mathbb{1}_{\leq \tau_j}, \phi \rangle_{\mathcal{H}} - \langle \Phi^n, \phi \rangle_{\mathcal{H}} \right\|_{L^2(\Omega \times [0,t]; \mathbb{R})}$$

$$= \left\| \langle \Psi.\mathbb{1}_{\leq \tau_j} - \Phi^n, \phi \rangle_{\mathcal{H}} \right\|_{L^2(\Omega \times [0,t]; \mathbb{R})}$$

$$\leq \left\| \|\Psi.\mathbb{1}_{\leq \tau_j} - \Phi^n\|_{\mathcal{H}} \|\phi\|_{\mathcal{H}} \right\|_{L^2(\Omega \times [0,t]; \mathbb{R})}$$

$$\leq \|\phi\|_{L^{\infty}(\Omega; \mathcal{H})} \left\| \|\Psi.\mathbb{1}_{\leq \tau_j} - \Phi^n\|_{\mathcal{H}} \right\|_{L^2(\Omega \times [0,t]; \mathbb{R})},$$

and by assumption,

$$\left\| \|\Psi.\mathbb{1}_{\leq \tau_j} - \Phi^n\|_{\mathcal{H}} \right\|_{L^2(\Omega \times [0,t]; \mathbb{R})} \longrightarrow 0$$

as $n \to \infty$, so the convergence is proved. To show the identity (2.13) in the case of the simple process Φ^n, observe that

$$\left\langle \int_0^t \Phi_r^n dW_r, \phi \right\rangle_{\mathcal{H}} = \left\langle \sum_{i=0}^{\infty} a_i^n \left(W_{t_{i+1}^n \wedge t} - W_{t_i^n \wedge t} \right), \phi \right\rangle_{\mathcal{H}}$$

$$= \sum_{i=0}^{\infty} \left\langle a_i^n \left(W_{t_{i+1}^n \wedge t} - W_{t_i^n \wedge t} \right), \phi \right\rangle_{\mathcal{H}}$$

$$= \sum_{i=0}^{\infty} \langle a_i^n, \phi \rangle_{\mathcal{H}} \left(W_{t_{i+1}^n \wedge t} - W_{t_i^n \wedge t} \right)$$

$$= \int_0^t \langle \Phi_r^n, \phi \rangle_{\mathcal{H}} dW_r$$

as required. In order to conclude the argument, by definition of the integral, we have that

$$\int_0^t \Psi_r \mathbb{1}_{r \leq \tau^j} dW_r = \lim_{n \to \infty} \int_0^t \Phi_r^n dW_r$$

$$\int_0^t \langle \Psi_r \mathbb{1}_{r \leq \tau^j}, \phi \rangle_{\mathcal{H}} dW_r = \lim_{n \to \infty} \int_0^t \langle \Phi_r^n, \phi \rangle_{\mathcal{H}} dW_r,$$

where the first limit is taken in $L^2(\Omega; \mathcal{H})$ and the second one in $L^2(\Omega; \mathbb{R})$. For each we can thus extract a $\mathbb{P} - a.s.$ convergent subsequence in the appropriate space, so by taking successive subsequences we can find one common subsequence indexed by (n_k) such that the above limits hold $\mathbb{P} - a.s..$ Thus

$$\left\langle \int_0^t \Psi_r \mathbb{1}_{r \leq \tau^j} dW_r, \phi \right\rangle_{\mathcal{H}} = \left\langle \lim_{n_k \to \infty} \int_0^t \Phi_r^{n_k} dW_r, \phi \right\rangle_{\mathcal{H}}$$

$$= \lim_{n_k \to \infty} \left\langle \int_0^t \Phi_r^{n_k} dW_r, \phi \right\rangle_{\mathcal{H}}$$

$$= \lim_{n_k \to \infty} \int_0^t \langle \Phi_r^n, \phi \rangle_{\mathcal{H}} dW_r$$

$$= \int_0^t \langle \Psi_r \mathbb{1}_{r \leq \tau^j}, \phi \rangle_{\mathcal{H}} dW_r$$

so (2.14) is justified and the proof is complete.

$\square$

We note that the $\mathcal{F}_0$-measurability requirement on ϕ really comes into play in showing the $\mathcal{F}_{t_i}$-measurability of $\langle a_i^n, \phi \rangle_{\mathcal{H}}$. If we were to consider the integral over some $[s, t]$ interval instead, then one could relax ϕ to only being $\mathcal{F}_s$-measurable. In fact the result can also be extended to unbounded ϕ. To do this we shall prove a Stochastic Dominated Convergence Theorem, Lemma 2.1.

Lemma 2.1 *Let* (Ψ^n) *be a sequence in* $\bar{\mathcal{I}}^{\mathcal{H}}$ *such that there exist processes* $\Psi :$ $\Omega \times [0, \infty) \to \mathcal{H}$ *and* $\Phi \in L^2([0, \infty); \mathbb{R})$ $\mathbb{P} - a.s.$, *with the properties that for every* $T > 0$, $\mathbb{P} \times \lambda - a.e.$ $(\omega, t) \in \Omega \times [0, T]$:

1. $\|\Psi_t^n(\omega)\|_{\mathcal{H}} \leq |\Phi_t(\omega)|$ *for all* $n \in \mathbb{N}$.
2. $(\Psi_t^n(\omega))$ *is convergent to* $\Psi_t(\omega)$ *in* $\mathcal{H}$.

Then $\boldsymbol{\Psi} \in \bar{I}^{\mathcal{H}}$, and for every $t > 0$, there exists a subsequence indexed by (n_k) such that

$$\lim_{n_k \to \infty} \int_0^t \boldsymbol{\Psi}_r^{n_k} dW_r = \int_0^t \boldsymbol{\Psi}_r dW_r \tag{2.15}$$

$\mathbb{P} - a.s..$

Proof Immediately, we note that $\boldsymbol{\Psi}$ inherits the progressive measurability from $\boldsymbol{\Psi}^n$ from the almost everywhere limit in the product space $\Omega \times [0, T]$ when equipped with product sigma algebra $\mathcal{F}_T \times \mathcal{B}([0, T])$. Similarly we must have that for $\mathbb{P} \times \lambda - a.e.$ (ω, t), $\|\boldsymbol{\Psi}_t(\omega)\|_{\mathcal{H}} \le |\boldsymbol{\Phi}_t(\omega)|$ so $\boldsymbol{\Psi} \in L^2([0, \infty); \mathcal{H})$ $\mathbb{P} - a.s.$ hence belongs to $\bar{I}^{\mathcal{H}}$. We look to find a common sequence of localizing times for the stochastic integrals and then demonstrate (2.15) by showing the identity holds true when stopped at each localizing time. To this end we introduce the stopping times

$$\tau_j := j \wedge \inf \left\{ 0 \le t < \infty : \int_0^t |\boldsymbol{\Phi}_r|^2 dr \ge j \right\},$$

which from the assumption 1 serve as a sequence of localizing times for every $\boldsymbol{\Psi}^n$, and too for $\boldsymbol{\Psi}$. Thus for any fixed $t > 0$ and $j \in \mathbb{N}$, we wish to show that

$$\lim_{n_k \to \infty} \int_0^t \boldsymbol{\Psi}_r^{n_k} \mathbb{1}_{r \le \tau_j} dW_r = \int_0^t \boldsymbol{\Psi}_r \mathbb{1}_{r \le \tau_j} dW_r \tag{2.16}$$

for a subsequence (n_k) $\mathbb{P} - a.s.$, or equivalently that

$$\lim_{n_k \to \infty} \int_0^t (\boldsymbol{\Psi}_r^{n_k} - \boldsymbol{\Psi}_r) \mathbb{1}_{r \le \tau_j} dW_r = 0.$$

We first assess the convergence in $L^2(\Omega; \mathcal{H})$, applying Corollary 2.1 for each fixed n to see that

$$\mathbb{E} \left\| \int_0^t (\boldsymbol{\Psi}_r^n - \boldsymbol{\Psi}_r) \mathbb{1}_{r \le \tau_j} dW_r \right\|_{\mathcal{H}}^2 = \mathbb{E} \left(\int_0^t \|(\boldsymbol{\Psi}_r^n - \boldsymbol{\Psi}_r) \mathbb{1}_{r \le \tau_j}\|_{\mathcal{H}}^2 dr \right).$$

Observing that for $\mathbb{P} \times \lambda - a.e.$ (ω, t),

$$\|(\boldsymbol{\Psi}_r^n(\omega) - \boldsymbol{\Psi}_r(\omega)) \mathbb{1}_{r \le \tau_j}(\omega)\|_{\mathcal{H}}^2 \le \left(\|\boldsymbol{\Psi}_r^n(\omega) \mathbb{1}_{r \le \tau_j}(\omega)\|_{\mathcal{H}} + \|\boldsymbol{\Psi}_r(\omega) \mathbb{1}_{r \le \tau_j}(\omega)\|_{\mathcal{H}} \right)^2$$

$$\le 4|\boldsymbol{\Phi}_r(\omega) \mathbb{1}_{r \le \tau_j}(\omega)|^2.$$

Then with dominating function $4|\boldsymbol{\Phi}.\mathbb{1}_{.\le \tau_j}|^2$, we can apply the standard Dominated Convergence Theorem for the integral over the product space (we face no problems with the order of integration from Tonelli's theorem given the progressive measurability) to deduce that

$$\lim_{n \to \infty} \mathbb{E} \left(\int_0^t \| (\boldsymbol{\Psi}_r^n - \boldsymbol{\Psi}_r) \mathbb{1}_{r \le \tau_j} \|_{\mathcal{H}}^2 dr \right) = 0,$$

and therefore

$$\lim_{n \to \infty} \mathbb{E} \left\| \int_0^t (\boldsymbol{\Psi}_r^n - \boldsymbol{\Psi}_r) \mathbb{1}_{r \le \tau_j} dW_r \right\|_{\mathcal{H}}^2 = 0.$$

Thus we have justified the convergence (2.16) in the sense of $L^2(\Omega; \mathcal{H})$, from which we can deduce a $\mathbb{P} - a.s.$ convergent subsequence and the result is proved. $\qquad \square$

Proposition 2.5 *Let* $\boldsymbol{\Psi} \in \bar{\mathcal{I}}^{\mathcal{H}}$ *and* $\phi : \Omega \to \mathcal{H}$ *be* $\mathcal{F}_0$-*measurable. Then* $\langle \boldsymbol{\Psi}, \phi \rangle \in \bar{\mathcal{I}}^{\mathcal{H}}$, *and for every* $t > 0$ *we have that*

$$\left\langle \int_0^t \boldsymbol{\Psi}_r dW_r, \phi \right\rangle_{\mathcal{H}} = \int_0^t \langle \boldsymbol{\Psi}_r, \phi \rangle_{\mathcal{H}} dW_r \tag{2.17}$$

$\mathbb{P} - a.s..$

Proof A justification that $\langle \boldsymbol{\Psi}_r, \phi \rangle_{\mathcal{H}} \in \bar{\mathcal{I}}^{\mathbb{R}}$ is precisely as in Proposition 2.4. To apply this result, we rewrite ϕ in a trivial way as

$$\phi := \sum_{k=1}^{\infty} \phi \mathbb{1}_{k \le \|\phi\|_{\mathcal{H}} < k+1},$$

where the limit is taken $\mathbb{P} - a.s.$ in $\mathcal{H}$ (similarly to (2.1) this is just a finite sum at each fixed ω, or more precisely just a single element of the sum). Introducing the notation

$$\phi^n := \sum_{k=1}^{n} \phi \mathbb{1}_{k \le \|\phi\|_{\mathcal{H}} < k+1}$$

then clearly $\phi^n \in L^\infty(\Omega; \mathcal{H})$ and is still $\mathcal{F}_0$-measurable, so we can apply Proposition 2.4 to see that

$$\left\langle \int_0^t \boldsymbol{\Psi}_r dW_r, \phi^n \right\rangle_{\mathcal{H}} = \int_0^t \langle \boldsymbol{\Psi}_r, \phi^n \rangle_{\mathcal{H}} dW_r.$$

We can take the $\mathbb{P} - a.s.$ limit in $\mathcal{H}$ outside of the inner product on the left hand side, so it is sufficient to show that

$$\lim_{n \to \infty} \int_0^t \langle \Psi_r, \phi^n \rangle_{\mathcal{H}} dW_r = \int_0^t \langle \Psi_r, \phi \rangle_{\mathcal{H}} dW_r \tag{2.18}$$

or at least that this is true for a subsequence. This is an immediate application of Lemma 2.1, with dominating function simply the limit $\langle \Psi, \phi \rangle_{\mathcal{H}}$. □

The same is true for multiplication by real valued random variables, where the proof is identical. We state the result here.

Proposition 2.6 *Let* $\Psi \in \bar{I}^{\mathcal{H}}$ *and* $\eta : \Omega \to \mathbb{R}$ *be* $\mathcal{F}_0$*-measurable. Then* $\eta\Psi \in \bar{I}^{\mathcal{H}}$, *and for every* $t > 0$ *we have that*

$$\eta \int_0^t \Psi_r dW_r = \int_0^t \eta \Psi_r dW_r$$

$\mathbb{P} - a.s..$

2.3 Cylindrical Brownian Motion

Having now addressed the question of how to integrate a Hilbert space valued process with respect to a finite dimensional local martingale, we look to extend this theory to the case of an infinite dimensional martingale. The aforementioned construction then arises from the one dimensional projections of the driving process. Our notion of infinite dimensional Brownian motion is that of a Cylindrical Brownian motion, which we explore in this section. Additional operator theory is used in this section to better understand the background of the process; however, it is not necessary in the resulting integral constructed in Sect. 2.5. We will denote by Q a bounded positive self-adjoint operator on $\mathcal{H}$, that is, $Q \in \mathscr{L}(\mathcal{H}; \mathcal{H})$, and $t \wedge s$ to be the minimum of $t, s \geq 0$.

Definition 2.12 A Q-*cylindrical Brownian motion* over $\mathcal{H}$ is a process W^Q taking values in the space of functions from $\mathcal{H}$ to $\mathbb{R}$, that is,

$$W^Q : \Omega \times [0, \infty) \to \mathbb{R}^{\mathcal{H}}$$

such that for each $h \in \mathcal{H}$ and $t \geq 0$, $W_h^Q(t) : \Omega \to \mathbb{R}$ is a zero mean Gaussian random variable, and for every $g, h \in \mathcal{H}$ and all times $t, s \geq 0$,

$$\mathbb{E}\big(W_g^Q(t) W_h^Q(s)\big) = \langle Qg, h \rangle_{\mathcal{H}}(t \wedge s).$$

Remark 2.2 We note that this definition is not restricted to a Brownian motion; one may consider any zero mean Gaussian random variable with a correlation function $R(t, s)$ replacing that of the Brownian motion $t \wedge s$. We consider only the Brownian motion case here to maintain a direct approach of the key principles, though the general case is defined in [52, Section 3.2.2].

To make this abstract definition workable, we look to see how such processes can be represented. This motivates the notion of a regular process.

Definition 2.13 A Q-cylindrical Brownian motion $\mathcal{W}^Q$ is said to be *regular* if there exists a square integrable $\mathcal{H}$-valued process $\widetilde{\mathcal{W}}^Q$ (that is, $\widetilde{\mathcal{W}}^Q_t \in L^2(\Omega; \mathcal{H})$ for all $t \geq 0$) such that for every $h \in \mathcal{H}$, $\mathcal{W}^Q_h$ has the same law as the process $\left\langle \widetilde{\mathcal{W}}^Q, h \right\rangle_{\mathcal{H}}$ defined by

$$\left\langle \widetilde{\mathcal{W}}^Q, h \right\rangle_{\mathcal{H}}(t, \omega) := \left\langle \widetilde{\mathcal{W}}^Q(t, \omega), h \right\rangle_{\mathcal{H}}.$$

We will not hesitate to identify the functional valued process $\mathcal{W}^Q$ with the Hilbert space one $\widetilde{\mathcal{W}}^Q$. In the next theorem however, we keep the distinction for clarity.

Theorem 2.3 *A Q-cylindrical Brownian motion $\mathcal{W}^Q$ is regular if and only if $Q \in \mathscr{L}^1(\mathcal{H}; \mathcal{H})$, that is, Q is trace-class. In this case $\mathcal{W}^Q$ admits the regular representation*

$$\widetilde{\mathcal{W}}^Q(t) = \sum_{i=1}^{\infty} \sqrt{\lambda_i} e_i W_t^i \tag{2.19}$$

for $t \geq 0$, where the elements (e_i) are an orthonormal basis of $\mathcal{H}$ consisting of eigenvectors of the self-adjoint trace-class (hence compact) operator Q, (λ_i) are the corresponding eigenvalues, and (W^i) are independent one dimensional Brownian motions. The infinite sum is taken as a limit in $L^2(\Omega; \mathcal{H})$.

Proof We consider the two implications. In the first one we show that if Q is trace-class, then (2.19) is a regular representation of $\mathcal{W}^Q$. In the second we show that if $\mathcal{W}^Q$ is regular, then Q is trace-class, so by the first implication, (2.19) is again a regular representation of $\mathcal{W}^Q$:

$\Longleftarrow$:   A sensible place to start would be to verify that for trace-class Q, (2.19) does indeed define an element of $L^2(\Omega; \mathcal{H})$. We will rely on completeness of the space and show that the sequence of partial sums is Cauchy. Observe that

$$\mathbb{E}\left(\left\| \sum_{i=m}^{n} \sqrt{\lambda_i} e_i W_t^i \right\|_{\mathcal{H}}^2\right) = \mathbb{E}\left(\sum_{i=m}^{n} |\lambda_i| |W_t^i|^2\right) = \mathbb{E}\left(|W_t|^2\right) \sum_{i=m}^{n} \lambda_i.$$

All we then require to conclude the Cauchy property is that $\sum_{i=1}^{\infty} \lambda_i < \infty$, which is given that Q is trace-class. To conclude that (2.19) is a regular representation of $\mathcal{W}^Q$, it is sufficient to show that the one dimensional processes $\left\langle \widetilde{\mathcal{W}}^Q, h \right\rangle_{\mathcal{H}}$ satisfy the conditions postulated of the $\mathcal{W}_h^Q$ in Definition 2.12 as these characterize the distribution.

First of all for each $h \in \mathcal{H}$ and time t, we must verify that the random variable

$$\omega \mapsto \left\langle \sum_{i=1}^{\infty} \sqrt{\lambda_i} e_i W_t^i(\omega), h \right\rangle_{\mathcal{H}} \tag{2.20}$$

is zero mean Gaussian. Note that

$$\left\langle \lim_{n \to \infty} \sum_{i=1}^{n} \sqrt{\lambda_i} e_i W_t^i, h \right\rangle_{\mathcal{H}} = \lim_{n \to \infty} \left\langle \sum_{i=1}^{n} \sqrt{\lambda_i} e_i W_t^i, h \right\rangle_{\mathcal{H}},$$

where the second limit is in $L^2(\Omega; \mathbb{R})$ just as we did in Theorem 2.1. The random variable (2.20) is thus an L^2 limit of zero mean Gaussian random variables, so is itself zero mean Gaussian (convergence in L^2 implies that in distribution so we have Gaussianity, and L^2 convergence implies L^1 from which we readily deduce the zero mean property). It remains to show that for each $g, h \in \mathcal{H}$ and $s, t > 0$ that

$$\mathbb{E}\left(\left\langle \sum_{i=1}^{\infty} \sqrt{\lambda_i} e_i W_t^i, g \right\rangle_{\mathcal{H}} \left\langle \sum_{j=1}^{\infty} \sqrt{\lambda_j} e_i W_s^j, h \right\rangle_{\mathcal{H}} \right) = \langle Qg, h \rangle_{\mathcal{H}} (t \wedge s).$$

We take the limit through the first inner product on the LHS as above, that is,

$$\mathbb{E}\left(\left(\lim_{n \to \infty} \left\langle \sum_{i=1}^{n} \sqrt{\lambda_i} e_i W_t^i, g \right\rangle_{\mathcal{H}} \right) \left\langle \sum_{j=1}^{\infty} \sqrt{\lambda_j} e_i W_s^j, h \right\rangle_{\mathcal{H}} \right), \tag{2.21}$$

and argue that for a sequence of functions (f_n) convergent to f in $L^2(\Omega; \mathbb{R})$, and $a \in L^2(\Omega; \mathbb{R})$, that

$$(\lim_{L^2} f_n)(a) = \lim_{L^1}(f_n a).$$

The right side is well defined as the limit of a Cauchy sequence:

$$\mathbb{E}\left(|f_n a - f_m a| \right) = \mathbb{E}\left(|(f_n - f_m)a| \right) \leq \mathbb{E}\left(|f_n - f_m|^2 \right) \mathbb{E}\left(|a|^2 \right),$$

and a similar calculation shows that this element of $L^1(\Omega; \mathbb{R})$ is the left side. Applying to (2.21) and pulling the L^1 limit through the expectation produce

$$\lim_{n \to \infty} \mathbb{E}\left(\left\langle \sum_{i=1}^{n} \sqrt{\lambda_i} e_i W_t^i, g \right\rangle_{\mathcal{H}} \left\langle \sum_{j=1}^{\infty} \sqrt{\lambda_j} e_j W_s^j, h \right\rangle_{\mathcal{H}} \right)$$

Playing the same game, this is

$$\lim_{n\to\infty}\lim_{m\to\infty}\mathbb{E}\left(\left\langle\sum_{i=1}^{n}\sqrt{\lambda_i}e_i\,W_t^i,g\right\rangle_{\mathcal{H}}\left\langle\sum_{j=1}^{m}\sqrt{\lambda_j}e_j\,W_s^j,h\right\rangle_{\mathcal{H}}\right)$$

and further

$$\lim_{n\to\infty}\lim_{m\to\infty}\mathbb{E}\left(\sum_{i=1}^{n}\sum_{j=1}^{m}\langle\sqrt{\lambda_i}e_i,g\rangle_{\mathcal{H}}\langle\sqrt{\lambda_j}e_j,h\rangle_{\mathcal{H}}W_t^i\,W_s^j\right).$$

Independence of the Brownian motions (W^i) and the fact that $t\wedge s$ is the correlation function of Brownian motion gives that this is equal to

$$\lim_{n\to\infty}\sum_{i=1}^{n}\langle\sqrt{\lambda_i}e_i,g\rangle_{\mathcal{H}}\langle\sqrt{\lambda_i}e_i,h\rangle_{\mathcal{H}}(t\wedge s)=\lim_{n\to\infty}\sum_{i=1}^{n}\langle\lambda_i e_i,g\rangle_{\mathcal{H}}\langle e_i,h\rangle_{\mathcal{H}}(t\wedge s),$$

which is just

$$\left\langle\sum_{i=1}^{\infty}\lambda_i\langle e_i,g\rangle e_i,h\right\rangle_{\mathcal{H}}(t\wedge s)=\langle Qg,h\rangle_{\mathcal{H}}(t\wedge s)$$

as required.

$\Longrightarrow$:   For the reverse direction, assume that $\mathcal{W}^Q$ is regular with corresponding process $\widetilde{\mathcal{W}}^Q$ (which is not necessarily of the form (2.19)). We want an expression in terms of the trace of Q, so we exploit Definition 2.12:

$$\sum_{i=1}^{\infty}\mathbb{E}\big(|\mathcal{W}_{e_i}^Q(t)|^2\big)=\sum_{i=1}^{\infty}\langle Qe_i,e_i\rangle_{\mathcal{H}}t,$$

and use an alternative expression from the assumed regular representation:

$$\sum_{i=1}^{\infty}\mathbb{E}\big(|\mathcal{W}_{e_i}^Q(t)|^2\big)=\sum_{i=1}^{\infty}\mathbb{E}\left(\left\langle\widetilde{\mathcal{W}}^Q(t),e_i\right\rangle_{\mathcal{H}}^2\right)$$

$$=\mathbb{E}\left(\sum_{i=1}^{\infty}\left\langle\widetilde{\mathcal{W}}^Q(t),e_i\right\rangle_{\mathcal{H}}^2\right)=\mathbb{E}\big(\big\|\widetilde{\mathcal{W}}^Q(t)\big\|_{\mathcal{H}}^2\big)<\infty,$$

where the infinite sum is pulled inside the expectation from the Monotone Convergence Theorem, and finiteness is by assumption on $\widetilde{\mathcal{W}}^Q\in L^2(\Omega;\mathcal{H})$. Hence Q is trace-class, and by the first implication, (2.19) is a regular representation of $\mathcal{W}^Q$.

$\square$

Remark 2.3 A standard cylindrical Brownian motion, that is, where Q is the identity, is thus not regular.

In light of Remark 2.3, it would be convenient to have such a representation for cylindrical Brownian motion (denoted simply by $\mathcal{W}$); we would like this to be along the lines of

$$\mathcal{W}_t = \sum_{i=1}^{\infty} e_i W_t^i, \tag{2.22}$$

where the (e_i) forms an orthonormal basis of $\mathcal{H}$, and (W^i) are real valued independent Brownian motions. We can in fact explicitly construct a larger Hilbert space $\mathcal{H}'$ such that the inclusion mapping $J : \mathcal{H} \hookrightarrow \mathcal{H}'$ is Hilbert–Schmidt. The composition $Q := JJ^*$ is then trace-class on $\mathcal{H}'$, and indeed $\mathcal{W}$ is a Q-cylindrical Brownian motion on $\mathcal{H}'$: We defer the details to, e.g., [52, Problem 3.2.6]. To be precise, that is $J(e_i) = \sqrt{\lambda_i}\eta_i$ for each i, where the (η_i) forms an orthonormal basis (of $\mathcal{H}'$) of eigenfunctions of JJ^* with eigenvalues λ_i, and for any $h \in \mathcal{H}$,

$$\mathcal{W}_h(t) = \left\langle \sum_{i=1}^{\infty} J(e_i)W_t^i, J(h) \right\rangle_{\mathcal{H}'}.$$

In this spirit, we consider (2.22) as a formal representation of cylindrical Brownian motion.

2.4   Martingale Theory in Hilbert Spaces

We first define the notion of a Hilbert space valued martingale, before addressing martingale properties of the stochastic integral.

Definition 2.14 A process M taking values in a Hilbert space $\mathcal{H}$ is said to be a martingale if for every $h \in \mathcal{H}$, the process $\langle M, h \rangle_{\mathcal{H}}$ is a real valued martingale. The martingale is said to be continuous if for $\mathbb{P} - a.e.$ ω and every $T > 0$, $M(\omega) : [0, T] \to \mathcal{H}$ is continuous. The martingale is said to be square integrable if for every $t \geq 0$, $\mathbb{E}(\|M_t\|_{\mathcal{H}}^2) < \infty$. The class of continuous square integrable martingales will be denoted $\mathcal{M}_c^2(\mathcal{H})$.[3]

We look to show that $\mathcal{M}_c^2(\mathcal{H})$ is closed in a suitable topology, for which we shall use a sufficient condition for continuity proven now.

---

[3] Recall that $\mathcal{M}_c^2 := \mathcal{M}_c^2(\mathbb{R})$, see Definition 2.7.

Lemma 2.2 *Let* $\psi \in C_w([0, T]; \mathcal{H})$ *and* $\|\psi\|_{\mathcal{H}}^2 \in C([0, T]; \mathbb{R})$, *that is,* ψ *is weakly continuous with continuous norm. Then* $\psi \in C([0, T]; \mathcal{H})$.

Proof We fix some $t \in [0, T]$ and look to show that ψ is continuous at t. To this end consider arbitrary $s \in [0, T]$. Then

$$
\begin{aligned}
\|\psi_t - \psi_s\|_{\mathcal{H}}^2 &= \langle \psi_t - \psi_s, \psi_t - \psi_s \rangle_{\mathcal{H}} \\
&= \langle \psi_t - \psi_s, \psi_t \rangle_{\mathcal{H}} - \langle \psi_t - \psi_s, \psi_s \rangle_{\mathcal{H}} \\
&= \langle \psi_t - \psi_s, \psi_t \rangle_{\mathcal{H}} + \|\psi_s\|_{\mathcal{H}}^2 - \langle \psi_t, \psi_s \rangle_{\mathcal{H}} \\
&= \langle \psi_t - \psi_s, \psi_t \rangle_{\mathcal{H}} + \|\psi_s\|_{\mathcal{H}}^2 + \langle \psi_t, \psi_t - \psi_s \rangle_{\mathcal{H}} - \|\psi_t\|_{\mathcal{H}}^2 \\
&= 2\langle \psi_t - \psi_s, \psi_t \rangle_{\mathcal{H}} + \left(\|\psi_s\|_{\mathcal{H}}^2 - \|\psi_t\|_{\mathcal{H}}^2 \right).
\end{aligned}
$$

For any given $\varepsilon > 0$, there exists a $\delta > 0$ such that for all $s \in [0 \vee (t-\delta), (t+\delta) \wedge T]$,

$$
\langle \psi_t - \psi_s, \psi_t \rangle_{\mathcal{H}} < \frac{\varepsilon}{3},
$$

$$
\|\psi_s\|_{\mathcal{H}}^2 - \|\psi_t\|_{\mathcal{H}}^2 < \frac{\varepsilon}{3}
$$

from each of the two assumptions, which gives the result. □

Proposition 2.7 *Let* (M^n) *be a sequence in* $\mathcal{M}_c^2(\mathcal{H})$ *and* M *a process with values in* $\mathcal{H}$ *such that at every time* $t \geq 0$,

$$
\lim_{n \to \infty} \mathbb{E}\left(\|M_t^n - M_t\|_{\mathcal{H}}^2 \right) = 0.
$$

Then $M \in \mathcal{M}_c^2(\mathcal{H})$.

Proof The square integrability is trivial, so we just consider the martingale property and continuity. For every $\phi \in \mathcal{H}$, at every $t \geq 0$, we have that

$$
\mathbb{E}\left(|\langle M_t^n, \phi \rangle_{\mathcal{H}} - \langle M_t, \phi \rangle_{\mathcal{H}}| \right) = \mathbb{E}\left(|\langle M_t^n - M_t, \phi \rangle_{\mathcal{H}}| \right) \leq \|\phi\|_{\mathcal{H}} \mathbb{E}\left(\|M_t^n - M_t\|_{\mathcal{H}} \right)
$$

$$
\leq \|\phi\|_{\mathcal{H}} \left(\mathbb{E}\left(\|M_t^n - M_t\|_{\mathcal{H}}^2 \right) \right)^{\frac{1}{2}}
$$

which converges to zero as $n \to \infty$. In particular, $\langle M_t^n, \phi \rangle_{\mathcal{H}}$ converges to $\langle M_t, \phi \rangle_{\mathcal{H}}$ in $L^1(\Omega; \mathbb{R})$, and as the former is a sequence of real valued martingales by definition, then it is standard that $\langle M, \phi \rangle_{\mathcal{H}}$ is again a martingale. As ϕ was arbitrary, we deduce the martingality of M. It remains to show pathwise continuity. For this we introduce the finite dimensional projections $(\mathcal{P}_k)$, defined by

$$\mathcal{P}_k : \phi \mapsto \sum_{i=1}^{k} \langle \phi, e_i \rangle_{\mathcal{H}} e_i$$

for (e_i) a fixed orthonormal basis of $\mathcal{H}$. The goal is to first show that $\mathcal{P}_k M$ has continuous paths for each k; as $\mathcal{P}_k$ is an orthogonal projection in $\mathcal{H}$, it is clear that at every $t \geq 0$,

$$\lim_{n \to \infty} \mathbb{E} \left(\left\| \mathcal{P}_k \left(M_t^n - M_t \right) \right\|_{\mathcal{H}}^2 \right) = 0,$$

and in particular

$$\lim_{n \to \infty} \mathbb{E} \left(\sum_{i=1}^{k} \langle M_t^n - M_t, e_i \rangle_{\mathcal{H}}^2 \right) = 0.$$

We can now exploit equivalent results from the finite dimensional theory: Note that

$$\lim_{n \to \infty} \mathbb{E} \left(\left| \sum_{i=1}^{k} \langle M_t^n, e_i \rangle - \sum_{i=1}^{k} \langle M_t, e_i \rangle_{\mathcal{H}} \right|^2 \right) \leq k \lim_{n \to \infty} \mathbb{E} \left(\sum_{i=1}^{k} \langle M_t^n - M_t, e_i \rangle_{\mathcal{H}}^2 \right) = 0,$$

so the sequence in n, $\left(\sum_{i=1}^{k} \langle M^n, e_i \rangle \right)$, of real valued martingales is convergent in $L^2(\Omega; \mathbb{R})$ at every $t \geq 0$. From Proposition A.1 we have that $\mathcal{M}_c^2$ is closed in this topology; hence $\sum_{i=1}^{k} \langle M, e_i \rangle$ is continuous. As this is true for every $k \in \mathbb{N}$, we may take differences to see that $\langle M, e_i \rangle$ is continuous for every i, and therefore $\mathcal{P}_k M$ is continuous in $\mathcal{H}$ for each k (of course, still pathwise $\mathbb{P} - a.s.$). In fact we take the same approach as the referenced proposition, working directly with the sequence of submartingales $\left(\sum_{i=1}^{k} \langle M, e_i \rangle_{\mathcal{H}}^2 \right)$ indexed by k. For any $\varepsilon > 0$, by Proposition A.2, we have that

$$\mathbb{P} \left(\left\{ \omega \in \Omega : \sup_{t \in [0,T]} \sum_{i=k+1}^{j} \langle M_t(\omega), e_i \rangle_{\mathcal{H}}^2 > \varepsilon \right\} \right) \leq \frac{1}{\varepsilon} \mathbb{E} \left(\sum_{i=k+1}^{j} \langle M_T, e_i \rangle_{\mathcal{H}}^2 \right),$$

(2.23)

but we have already established that $\mathbb{E} \left(\| M_T \|_{\mathcal{H}}^2 \right) < \infty$, or equivalently

$$\mathbb{E} \left(\sum_{i=1}^{\infty} \langle M_t(\omega), e_i \rangle_{\mathcal{H}}^2 \right) < \infty.$$

Therefore in the limit as $k \to \infty$, (2.23) approaches zero uniformly in $j > k$. One may choose a subsequence such that for all $k_l < k_m$,

$$\mathbb{P}\left(\left\{\omega \in \Omega: \sup_{t \in [0,T]} \sum_{i=k_l+1}^{k_m} \langle M_t(\omega), e_i \rangle_{\mathcal{H}}^2 > \frac{1}{l}\right\}\right) \le \frac{1}{2^l}.$$

By the Borel Cantelli Lemma the subsequence $\sum_{i=1}^{k_l} \langle M, e_i \rangle_{\mathcal{H}}^2$ is Cauchy in $C([0, T]; \mathbb{R})$ $\mathbb{P} - a.s..$ It thus admits a limit $\mathbb{P} - a.s.$ in $C([0, T]; \mathbb{R})$, which agrees with the limit at each t, given by $\sum_{i=1}^{\infty} \langle M_t, e_i \rangle_{\mathcal{H}}^2$, which is of course $\|M_t\|_{\mathcal{H}}^2$. Therefore, the process $\|M\|_{\mathcal{H}}^2$ is pathwise continuous; from Lemma 2.2, it is now sufficient to just show weak continuity. This is clear however, as for any given $\phi \in \mathcal{H}$ and $t \ge 0$, we have again that

$$\lim_{n \to \infty} \mathbb{E}\left(\langle M_t^n - M_t, \phi \rangle_{\mathcal{H}}^2\right) = 0,$$

so $\langle M, \phi \rangle_{\mathcal{H}}$ is shown to belong to $\mathcal{M}_c^2$ exactly as was done for $\sum_{i=1}^{k} \langle M, e_i \rangle_{\mathcal{H}}$, thus concluding the proof. $\square$

Proposition 2.8 *For a standard real valued Brownian motion W and $\mathbf{\Psi} \in I^{\mathcal{H}}$, the Itô stochastic integral*

$$\int_0^{\cdot} \mathbf{\Psi}_s dW_s$$

belongs to $\mathcal{M}_c^2(\mathcal{H})$.

Proof Recalling Definition 2.6, the integral is defined at each time t as a limit in $L^2(\Omega; \mathcal{H})$ of simple integrals. From Proposition 2.7 it is sufficient to show that these approximating integrals all belong to $\mathcal{M}_c^2(\mathcal{H})$, which is clear referring to the representation (2.1) and the definition of $\mathcal{M}_c^2(\mathcal{H})$. $\square$

Proposition 2.8 extends to the case of a general martingale integrator as in the finite dimensional setting. Local martingality is then defined as we would expect, and we have the following result.

Proposition 2.9 *For a continuous local martingale N and $\mathbf{\Psi} \in \bar{I}_N^{\mathcal{H}}$, the Itô stochastic integral*

$$\int_0^t \mathbf{\Psi}_s dN_s$$

is itself a continuous local martingale.

Proof We claim that the localizing stopping times are given simply by the (τ_n) as defined in (2.11). We have already seen that these tend to infinity almost surely, so it just remains to show that at any fixed $n \in \mathbb{N}$ the stopped process is a continuous martingale. The argument is now exactly as in Proposition 2.8. $\square$

The next ingredient would be a definition of quadratic variation, which we look to do via a Doob–Meyer decomposition for $M \in \mathcal{M}_c^2(\mathcal{H})$ (Theorem A.1). For this however, we must first define a notion of uniqueness in the decomposition.

Definition 2.15 Two $\mathcal{H}$-valued processes Ψ, Φ are said to be *indistinguishable* if there exists a set $A \in \mathcal{F}$ with $\mathbb{P}(A) = 1$ such that at all $t \geq 0$ and all $\omega \in A$, $\Psi_t(\omega) = \Phi_t(\omega)$.

As we have become accustomed to, this notion can be built up from the finite dimensional projections and one dimensional theory.

Lemma 2.3 *Let Ψ and Φ be $\mathcal{H}$-valued processes. Then Ψ is indistinguishable from Φ if and only if for every basis vector e_i, $\langle \Psi, e_i \rangle_{\mathcal{H}}$ is indistinguishable from $\langle \Phi, e_i \rangle_{\mathcal{H}}$.*

Proof The first implication is trivial so we consider only the reverse one. That is, assume that for every e_i there exists a set $A_i \in \mathcal{F}$ with $\mathbb{P}(A_i) = 1$, and for all $t \geq 0$ and $\omega \in A_i$,

$$\langle \Psi_t(\omega), e_i \rangle_{\mathcal{H}} = \langle \Phi_t(\omega), e_i \rangle_{\mathcal{H}}.$$

We now define $A := \bigcap_i A_i$, which is again of full probability and in $\mathcal{F}$, and for any $\omega \in A, t \geq 0$,

$$\|\Psi_t(\omega) - \Phi_t(\omega)\|_{\mathcal{H}}^2 = \sum_{i=1}^{\infty} \langle \Psi_t(\omega) - \Phi_t(\omega), e_i \rangle_{\mathcal{H}}^2 = 0,$$

which completes the proof. □

Revisiting the quadratic variation, our question is: Can we show that $\|M\|_{\mathcal{H}}^2$ defines a submartingale? We have integrability by definition, and

$$\|M_t\|_{\mathcal{H}}^2 = \sum_{i=1}^{\infty} \langle M_t, e_i \rangle_{\mathcal{H}}^2,$$

where the limit is defined $\mathbb{P} - a.s.$. Again by definition the above projections are martingales and so the squares are submartingales. The process $\|M\|_{\mathcal{H}}^2$ is adapted as each $\|M_t\|_{\mathcal{H}}^2$ is the $\mathbb{P} - a.s.$ limit of $\mathcal{F}_t$-measurable random variables (on the complete measure space), and it is a submartingale as we can apply the Monotone Convergence Theorem to take the limit through the expectation for the defining submartingale property. As such, we have the following:

Definition 2.16 For $M \in \mathcal{M}_c^2(\mathcal{H})$, the quadratic variation $[M]$ of M is defined to be the unique[4], continuous, adapted, nondecreasing process with $[M]_0 = 0$ ($\mathbb{P} - a.s.$)

---

[4] Uniqueness here is "up to indistinguishability," as defined in Definition 2.15.

specified in the Doob–Meyer decomposition (Theorem A.1) such that

$$\|M\|^2_{\mathcal{H}} - [M]$$

is a real valued martingale.

Proposition 2.10 *Suppose that* $\boldsymbol{\Psi} \in I^{\mathcal{H}}$, *then*

$$\left[\int_0^{\cdot} \boldsymbol{\Psi}_r dW_r\right]_t = \int_0^t \|\boldsymbol{\Psi}_r\|^2_H dr. \qquad (2.24)$$

Proof The fact that the process in (2.24) is continuous, adapted, nondecreasing, and starting from zero is clear. It simply remains to show the required martingality. To this end observe that at each time t,

$$\left\|\int_0^t \boldsymbol{\Psi}_r dW_r\right\|^2_{\mathcal{H}} - \int_0^t \|\boldsymbol{\Psi}_r\|^2_H dr$$

$$= \sum_{i=1}^{\infty} \left\langle \int_0^t \boldsymbol{\Psi}_r dW_r, e_i \right\rangle^2_{\mathcal{H}} - \int_0^t \sum_{i=1}^{\infty} \langle \boldsymbol{\Psi}_r, e_i \rangle^2_H dr$$

$$= \sum_{i=1}^{\infty} \left(\int_0^t \langle \boldsymbol{\Psi}_r, e_i \rangle_{\mathcal{H}} dW_r\right)^2 - \sum_{i=1}^{\infty} \int_0^t \langle \boldsymbol{\Psi}_r, e_i \rangle^2_H dr$$

$$= \sum_{i=1}^{\infty} \left(\left(\int_0^t \langle \boldsymbol{\Psi}_r, e_i \rangle_{\mathcal{H}} dW_r\right)^2 - \int_0^t \langle \boldsymbol{\Psi}_r, e_i \rangle^2_H dr\right)$$

having applied Theorem 2.1 to the first term and the Monotone Convergence Theorem to the second term, where the infinite sum is a limit taken $\mathbb{P} - a.s..$ From the standard one dimensional theory, for each n,

$$\sum_{i=1}^{n} \left(\left(\int_0^t \langle \boldsymbol{\Psi}_r, e_i \rangle_{\mathcal{H}} dW_r\right)^2 - \int_0^t \langle \boldsymbol{\Psi}_r, e_i \rangle^2_H dr\right)$$

is a real valued martingale so to conclude the proof we only need to justify that the $\mathbb{P} - a.e.$ limit also holds in $L^1(\Omega; \mathbb{R})$ as convergence in this space preserves martingality. This is a straightforward application of the Monotone Convergence Theorem applied to each integral separately, which concludes the proof.   □

We also look to reconcile this definition with the one often stated in the one dimensional case, as a limit in probability over any time partition with mesh approaching zero.

Proposition 2.11 *Let* $\boldsymbol{\Psi} \in I^{\mathcal{H}}_T$ *and consider any sequence of partitions*

$$I_l := \left\{ 0 = t_0^l < t_1^l < \cdots < t_{k_l}^l = T \right\}$$

with $\max_j |t_j^l - t_{j-1}^l| \to 0$ as $l \to \infty$. Then for all $t \in [0, T]$, for any $\varepsilon > 0$,

$$\lim_{l \to \infty} \mathbb{P}\left(\left\{ \left| \sum_{t_{j+1}^l \leq t} \left\| \int_{t_j^l}^{t_{j+1}^l} \boldsymbol{\Psi}_r dW_r \right\|_{\mathcal{H}}^2 - \int_0^t \|\boldsymbol{\Psi}_r\|_{\mathcal{H}}^2 dr \right| > \varepsilon \right\} \right) = 0.$$

We prove this result with an intermediary lemma, stated in the setting of Proposition 2.11.

Lemma 2.4 *Define the sequence of stopping times* (τ^n) *at every* $n \in \mathbb{N}$ *by*

$$\tau^n := n \wedge \inf \left\{ t \in [0, T] : \left\| \int_0^t \boldsymbol{\Psi}_r dW_r \right\|_{\mathcal{H}}^2 + \int_0^t \|\boldsymbol{\Psi}_r\|_{\mathcal{H}}^2 dr \geq n \right\}$$

and the process $\boldsymbol{\Psi}^n := \boldsymbol{\Psi} . \mathbb{1}_{\cdot \leq \tau^n}$. *Suppose that for every* n *and all* $t \in [0, T]$,

$$\lim_{l \to \infty} \mathbb{E}\left(\left| \sum_{t_{j+1}^l \leq t} \left\| \int_{t_j^l}^{t_{j+1}^l} \boldsymbol{\Psi}_r^n dW_r \right\|_{\mathcal{H}}^2 - \int_0^t \|\boldsymbol{\Psi}_r^n\|_{\mathcal{H}}^2 dr \right| \right) = 0. \tag{2.25}$$

Then for any $\varepsilon > 0$,

$$\lim_{l \to \infty} \mathbb{P}\left(\left\{ \left| \sum_{t_{j+1}^l \leq t} \left\| \int_{t_j^l}^{t_{j+1}^l} \boldsymbol{\Psi}_r dW_r \right\|_{\mathcal{H}}^2 - \int_0^t \|\boldsymbol{\Psi}_r\|_{\mathcal{H}}^2 dr \right| > \varepsilon \right\} \right) = 0.$$

Proof Define

$$A^l := \left\{ \left| \sum_{t_{j+1}^l \leq t} \left\| \int_{t_j^l}^{t_{j+1}^l} \boldsymbol{\Psi}_r dW_r \right\|_{\mathcal{H}}^2 - \int_0^t \|\boldsymbol{\Psi}_r\|_{\mathcal{H}}^2 dr \right| > \varepsilon \right\}$$

and then for any n,

$$A^l = A^l \cap \left[\{\tau^n > T\} \cup \{\tau^n \leq T\} \right] = \left[A^l \cap \{\tau^n > T\} \right] \cup \left[A^l \cap \{\tau^n \leq T\} \right]$$

$$\subset \left[A^l \cap \{\tau^n > T\} \right] \cup \{\tau^n \leq T\}.$$

In particular,

$$\mathbb{P}\left(A^l\right) \leq \mathbb{P}\left(A^l \cap \{\tau^n > T\}\right) + \mathbb{P}\left(\{\tau^n \leq T\}\right), \tag{2.26}$$

where

$$A^{l,n} := A^l \cap \{\tau^n > T\} = \left\{ \left| \sum_{t^l_{j+1} \leq t} \left\| \int_{t^l_j}^{t^l_{j+1}} \boldsymbol{\Psi}^n_r dW_r \right\|^2_{\mathcal{H}} - \int_0^t \|\boldsymbol{\Psi}^n_r\|^2_{\mathcal{H}} dr \right| > \varepsilon \right\}.$$

We then take the limit as $l \to \infty$ in (2.26), which holds for all n so it must hold in the limit as $n \to \infty$, providing that

$$\lim_{l\to\infty} \mathbb{P}\left(A^l\right) \leq \lim_{n\to\infty}\lim_{l\to\infty} \mathbb{P}\left(A^l \cap \{\tau^n > T\}\right) + \lim_{n\to\infty} \mathbb{P}\left(\{\tau^n \leq T\}\right)$$

$$= \lim_{n\to\infty}\lim_{l\to\infty} \mathbb{P}\left(A^l \cap \{\tau^n > T\}\right) \tag{2.27}$$

given that the (τ^n) approach infinity $\mathbb{P} - a.s..$ From Chebyshev's inequality and the assumption (2.25),

$$\lim_{n\to\infty}\lim_{l\to\infty} \mathbb{P}\left(A^l \cap \{\tau^n > T\}\right)$$

$$\leq \frac{1}{\varepsilon} \lim_{n\to\infty}\lim_{l\to\infty} \mathbb{E}\left(\left| \sum_{t^l_{j+1} \leq t} \left\| \int_{t^l_j}^{t^l_{j+1}} \boldsymbol{\Psi}^n_r dW_r \right\|^2_{\mathcal{H}} - \int_0^t \|\boldsymbol{\Psi}^n_r\|^2_{\mathcal{H}} dr \right|\right) = 0,$$

which combined with (2.27) gives the result.   □

Proof of Proposition 2.11 We once again look to prove this by considering the finite dimensional projections on which the result is known to be true, before showing that it is preserved in the limit. This approach is ultimately taken for the localized sequence by applying Lemma 2.4, looking to verify (2.25). Identically to the proof of Proposition 2.10, we have that

$$\mathbb{E}\left(\left| \sum_{t^l_{j+1} \leq t} \left\| \int_{t^l_j}^{t^l_{j+1}} \boldsymbol{\Psi}^n_r dW_r \right\|^2_{\mathcal{H}} - \int_0^t \|\boldsymbol{\Psi}^n_r\|^2_H dr \right|\right)$$

$$= \mathbb{E}\left(\left| \sum_{t^l_{j+1} \leq t} \sum_{i=1}^{\infty} \left(\int_{t^l_j}^{t^l_{j+1}} \langle\boldsymbol{\Psi}^n_r, e_i\rangle_{\mathcal{H}} dW_r \right)^2 - \int_0^t \langle\boldsymbol{\Psi}^n_r, e_i\rangle^2_H dr \right|\right)$$

$$\leq \mathbb{E}\left(\sum_{i=1}^{\infty}\left|\sum_{t_{j+1}^{l}\leq t}\left(\int_{t_j^l}^{t_{j+1}^l}\langle\boldsymbol{\Psi}_r^n,e_i\rangle_{\mathcal{H}}dW_r\right)^2 - \int_0^t\langle\boldsymbol{\Psi}_r^n,e_i\rangle_H^2 dr\right|\right)$$

$$= \sum_{i=1}^{\infty}\mathbb{E}\left(\left|\sum_{t_{j+1}^{l}\leq t}\left(\int_{t_j^l}^{t_{j+1}^l}\langle\boldsymbol{\Psi}_r^n,e_i\rangle_{\mathcal{H}}dW_r\right)^2 - \int_0^t\langle\boldsymbol{\Psi}_r^n,e_i\rangle_H^2 dr\right|\right),$$

where on the last line we have applied the Monotone Convergence Theorem. From Theorem A.3 we know that for each $i \in \mathbb{N}$,

$$\lim_{l\to 0}\mathbb{E}\left(\left|\sum_{t_{j+1}^{l}\leq t}\left[\left(\int_{t_j^l}^{t_{j+1}^l}\langle\boldsymbol{\Psi}_r^n,e_i\rangle_{\mathcal{H}}dW_r\right)^2 - \int_0^t\langle\boldsymbol{\Psi}_r^n,e_i\rangle_H^2 dr\right]\right|\right) = 0,$$

so we would be done if we can justify the interchange of infinite sum and limit in l. We look to apply the Dominated Convergence Theorem to proceed, noting that for each fixed $i, l \in \mathbb{N}$,

$$\mathbb{E}\left(\left|\sum_{t_{j+1}^{l}\leq t}\left[\left(\int_{t_j^l}^{t_{j+1}^l}\langle\boldsymbol{\Psi}_r^n,e_i\rangle_{\mathcal{H}}dW_r\right)^2 - \int_0^t\langle\boldsymbol{\Psi}_r^n,e_i\rangle_H^2 dr\right]\right|\right)$$

$$\leq \sum_{t_{j+1}^{l}\leq t}\mathbb{E}\left[\left(\int_{t_j^l}^{t_{j+1}^l}\langle\boldsymbol{\Psi}_r^n,e_i\rangle_{\mathcal{H}}dW_r\right)^2\right] + \mathbb{E}\left[\int_0^t\langle\boldsymbol{\Psi}_r^n,e_i\rangle_H^2 dr\right]$$

$$= \sum_{t_{j+1}^{l}\leq t}\mathbb{E}\left[\int_{t_j^l}^{t_{j+1}^l}\langle\boldsymbol{\Psi}_r^n,e_i\rangle_H^2 dr\right] + \mathbb{E}\left[\int_0^t\langle\boldsymbol{\Psi}_r^n,e_i\rangle_H^2 dr\right]$$

$$= 2\mathbb{E}\left[\int_0^t\langle\boldsymbol{\Psi}_r^n,e_i\rangle_H^2 dr\right], \tag{2.28}$$

which is a bound uniform in l and summable in i. Thus, $2\mathbb{E}\left[\int_0^t\langle\boldsymbol{\Psi}_r^n,e_i\rangle_H^2 dr\right]$ is our dominating function which justifies the application of the Dominated Convergence Theorem, concluding the verification of (2.25) and hence the proof by Lemma 2.4. $\qquad\square$

Lemma 2.5 *Suppose that (M^n) is a sequence of martingales in $\mathcal{M}_c^2(\mathcal{H})$ which at every time $t \geq 0$ converges in $L^2(\Omega;\mathcal{H})$ to some M_t. Suppose in addition that at any time $t \geq 0$, the sequence $([M^n]_t)$ converges to some ψ_t in $L^1(\Omega;\mathbb{R})$ where ψ is continuous, adapted, and nondecreasing ($\mathbb{P}-a.s.$). Then $M \in \mathcal{M}_c^2(\mathcal{H})$ and $[M]$ is indistinguishable from ψ.*

Proof The fact that $M \in \mathcal{M}_c^2(\mathcal{H})$ is immediate from Proposition 2.7, so we move on to the quadratic variation. Observe that for each n, by definition

$$\|M^n\|_{\mathcal{H}}^2 - [M^n]$$

is a real valued martingale, so the $L^1(\Omega; \mathbb{R})$ limit at each time t,

$$\lim_{n \to \infty} \left(\|M^n\|_{\mathcal{H}}^2 - [M^n] \right)$$

is again a real valued martingale if it exists. But this is clear as we have that

$$\lim_{n \to \infty} \|M^n\|_{\mathcal{H}}^2 - \lim_{n \to \infty} [M^n] = \|M\|_{\mathcal{H}}^2 - \psi$$

by definition of the limit. From Definition 2.16, then if $\psi_0 = 0 \, \mathbb{P}-a.s.$, we have that ψ satisfies the conditions of the quadratic variation of M, so it is indistinguishable from it. This property is immediate as ψ_0 is the limit in $L^1(\Omega; \mathbb{R})$ of $([M^n]_0)$ by assumption, while this is just a sequence of zeros $\mathbb{P} - a.s.$, hence the result. $\square$

In part due to this result, we provide an alternative characterization of the quadratic variation under some additional boundedness assumptions. We reintroduce the projections $(\mathcal{P}_k)$ used in Proposition 2.7.

Proposition 2.12 *For any $M \in \mathcal{M}_c^2(\mathcal{H})$, $[M]$ has the representation*

$$[M] = \sum_{i=1}^{\infty} \left[\langle M, e_i \rangle_{\mathcal{H}} \right]$$

$\mathbb{P} - a.s.$ *for the limit taken in $C([0, T]; \mathbb{R})$ for any $T \geq 0$.*

Proof Observe that

$$\|\mathcal{P}_k M\|_{\mathcal{H}}^2 = \sum_{i=1}^{k} \langle M, e_i \rangle_{\mathcal{H}}^2,$$

which is the sum of k one dimensional submartingales, and in particular

$$\sum_{i=1}^{k} \left(\langle M, e_i \rangle_{\mathcal{H}}^2 - [\langle M, e_i \rangle_{\mathcal{H}}] \right)$$

is a martingale. From the definition of the quadratic variation, it is thus clear that

$$[\mathcal{P}_k M] = \sum_{i=1}^{k} \left[\langle M, e_i \rangle_{\mathcal{H}} \right]. \tag{2.29}$$

Moreover we have that at each $t \geq 0$, $\mathcal{P}_k M_t$ is convergent to M_t $\mathbb{P} - a.s.$ in $\mathcal{H}$, and furthermore in $L^2(\Omega; \mathcal{H})$ from an application of the Monotone Convergence Theorem to the difference process $\|M - \mathcal{P}_k M\|_{\mathcal{H}}^2 = \sum_{i=k+1}^{\infty} \langle M, e_i \rangle_{\mathcal{H}}^2$. Reminiscent of Lemma 2.5, we look to show convergence of the sequence $([\mathcal{P}_k M])$ to some ψ at each $t \geq 0$ in $L^1(\Omega; \mathbb{R})$, and in fact we show the stronger convergence in $L^1(\Omega; C([0, T]; \mathbb{R}))$ for each $T \geq 0$. We proceed by showing the Cauchy property in this Banach Space. To do this we consider, for $j < k$,

$$[\mathcal{P}_k M] - [\mathcal{P}_j M] = \sum_{i=j+1}^{k} \left[\langle M, e_i \rangle_{\mathcal{H}} \right]. \tag{2.30}$$

From this identity and the preceding work, it is clear that

$$[\mathcal{P}_k M] - [\mathcal{P}_j M] = [\mathcal{P}_k M - \mathcal{P}_j M] \tag{2.31}$$

and therefore

$$\sup_{t \in [0,T]} \left| [\mathcal{P}_k M]_t - [\mathcal{P}_j M]_t \right| = \sup_{t \in [0,T]} [\mathcal{P}_k M - \mathcal{P}_j M]_t = [\mathcal{P}_k M - \mathcal{P}_j M]_T. \tag{2.32}$$

Furthermore, we are concerned with a control in expectation of this term. From the property that real valued martingales have constant expectation, we deduce that

$$\mathbb{E} \left(\left\| \mathcal{P}_k M_T - \mathcal{P}_j M_T \right\|_{\mathcal{H}}^2 - [\mathcal{P}_k M - \mathcal{P}_j M]_T \right)$$

$$= \mathbb{E} \left(\left\| \mathcal{P}_k M_0 - \mathcal{P}_j M_0 \right\|_{\mathcal{H}}^2 - [\mathcal{P}_k M - \mathcal{P}_j M]_0 \right)$$

$$= \mathbb{E} \left(\left\| \mathcal{P}_k M_0 - \mathcal{P}_j M_0 \right\|_{\mathcal{H}}^2 \right)$$

so in particular

$$\mathbb{E} \left([\mathcal{P}_k M - \mathcal{P}_j M]_T \right) = \mathbb{E} \left(\left\| \mathcal{P}_k M_T - \mathcal{P}_j M_T \right\|_{\mathcal{H}}^2 - \left\| \mathcal{P}_k M_0 - \mathcal{P}_j M_0 \right\|_{\mathcal{H}}^2 \right)$$

$$= \mathbb{E} \left(\sum_{i=j+1}^{k} \left(\langle M_T, e_i \rangle_{\mathcal{H}}^2 - \langle M_0, e_i \rangle_{\mathcal{H}}^2 \right) \right)$$

$$\leq \mathbb{E} \left(\sum_{i=j+1}^{k} \langle M_T, e_i \rangle_{\mathcal{H}}^2 \right) + \mathbb{E} \left(\sum_{i=j+1}^{k} \langle M_0, e_i \rangle_{\mathcal{H}}^2 \right).$$

By the square integrability assumption on M, we have that at every $t \geq 0$, $\mathbb{E}\left(\sum_{i=1}^{\infty} \langle M_t, e_i \rangle_{\mathcal{H}}^2\right) < \infty$, which justifies that the right hand side of the above approaches zero as $j \to \infty$, uniformly in k. With (2.32) we deduce that the sequence $([\mathcal{P}_k M])$ is Cauchy in $L^1(\Omega; C([0, T]; \mathbb{R}))$ for each $T \geq 0$ so it admits a limit in this space which we call ψ. Through the $L^1(\Omega; C([0, T]; \mathbb{R}))$ convergence, we can deduce the existence of a subsequence which is $\mathbb{P} - a.s.$ convergent in $C([0, T]; \mathbb{R})$. In fact we can upgrade this to convergence over the whole sequence, and as was noted that $\mathbb{P} - a.s.$ at each $t \geq 0$ the real valued sequence $([\mathcal{P}_k M]_t)$ not only has a convergent subsequence but is nondecreasing in k, which implies convergence of the whole sequence. Convergence of the whole sequence $\mathbb{P} - a.s.$ in $C([0, T]; \mathbb{R})$ can then be deduced by the Cauchy property with (2.32) and (2.31). In summary thus far we have that

$$\psi = \sum_{i=1}^{\infty} [\langle M, e_i \rangle_{\mathcal{H}}] \quad \mathbb{P} - a.s.$$

for the limit (of the sequence of finite sums) taken in $C([0, T]; \mathbb{R})$ for any $T \geq 0$. It only remains to show that ψ is indistinguishable from $[M]$, which we do by verifying the conditions of Lemma 2.5. The convergence has already been established; hence we need only the regularity on ψ. It is continuous by construction and must be adapted as ψ_t is the $\mathbb{P} - a.s.$ limit of the $\mathcal{F}_t$-measurable $[\mathcal{P}_k M]_t$ on the complete measure space. This limit similarly preserves the nondecreasing property, so ψ satisfies the required conditions and must be indistinguishable from $[M]$ due to Lemma 2.5.   □

Following on from the quadratic variation, we look to introduce the cross-variation between martingales. Guided by the motivation to use the Stratonovich integral, introduced in Sect. 3.1, we will need to consider the cross-variation between elements of $\mathcal{M}_c^2(\mathcal{H})$ and $\mathcal{M}_c^2$. Defining this akin to a polarization identity is out of the question as one cannot take sums of these martingales (there is no canonical way to sum an element of a Hilbert space with an element of $\mathbb{R}$), so we look to use the characterization in terms of their product.

Indeed from the classical theory, Theorem A.2, for any given $\boldsymbol{\Psi} \in \mathcal{M}_c^2(\mathcal{H})$, $Y \in \mathcal{M}_c^2$, and e_i a basis vector of $\mathcal{H}$, there exists a unique continuous, adapted, bounded-variation process $[\langle \boldsymbol{\Psi}, e_i \rangle_{\mathcal{H}}, Y]$ with $[\langle \boldsymbol{\Psi}, e_i \rangle_{\mathcal{H}}, Y]_0 = 0$ ($\mathbb{P} - a.s.$) such that

$$\langle \boldsymbol{\Psi}, e_i \rangle_{\mathcal{H}} Y - [\langle \boldsymbol{\Psi}, e_i \rangle_{\mathcal{H}}, Y]$$

is a real valued martingale. We would love to immediately have the existence and uniqueness of a corresponding $\mathcal{H}$-valued process which gives a martingale when subtracted from $\boldsymbol{\Psi} Y$, but that is not clear in the same way that the quadratic variation was as in that case $\|\boldsymbol{\Psi}\|_{\mathcal{H}}^2$ was a genuine real valued submartingale. Our approach, therefore, comes from the characterization in Proposition 2.12. We make a first definition for the projected process.

Definition 2.17 For $\Psi \in \mathcal{M}_c^2(\mathcal{H})$ and $Y \in \mathcal{M}_c^2$, for any $k \in \mathbb{N}$, we define the cross-variation process $[\mathcal{P}_k\Psi, Y]$ by

$$[\mathcal{P}_k\Psi, Y] := \sum_{i=1}^{k}[\langle\Psi, e_i\rangle_{\mathcal{H}}, Y]e_i.$$

Indeed we can verify that such a process gives us the desired properties of a cross-variation in $\mathcal{H}$.

Proposition 2.13 *For $\Psi \in \mathcal{M}_c^2(\mathcal{H})$ and $Y \in \mathcal{M}_c^2$, for any $k \in \mathbb{N}$, $[\mathcal{P}_k\Psi, Y]$ is the unique continuous, adapted, bounded-variation $\mathcal{H}$-valued process satisfying $[\mathcal{P}_k\Psi, Y]_0 = 0 \; \mathbb{P} - a.s.$ such that*

$$(\mathcal{P}_k\Psi)Y - [\mathcal{P}_k\Psi, Y]$$

is an $\mathcal{H}$-valued martingale.

Proof It follows from the corresponding properties of the real valued cross-variations that $[\mathcal{P}_k\Psi, Y]$ is continuous, adapted, and of bounded-variation (one can apply the triangle inequality for the norm $\|\cdot\|_{\mathcal{H}}$) satisfying $[\mathcal{P}_k\Psi, Y]_0 = 0 \; \mathbb{P} - a.s..$ In addition observe that

$$(\mathcal{P}_k\Psi)Y - [\mathcal{P}_k\Psi, Y] = \sum_{i=1}^{k}\left(\langle\Psi, e_i\rangle_{\mathcal{H}}Y - [\langle\Psi, e_i\rangle_{\mathcal{H}}, Y]\right)e_i,$$

which we look to show it is an $\mathcal{H}$-valued martingale. To this end consider arbitrary $\phi \in \mathcal{H}$. Then

$$\langle(\mathcal{P}_k\Psi)Y - [\mathcal{P}_k\Psi, Y], \phi\rangle_{\mathcal{H}}$$

$$= \left\langle \sum_{i=1}^{k}\left(\langle\Psi, e_i\rangle_{\mathcal{H}}Y - [\langle\Psi, e_i\rangle_{\mathcal{H}}, Y]\right)e_i, \sum_{j=1}^{\infty}\langle\phi, e_j\rangle_{\mathcal{H}}e_j \right\rangle_{\mathcal{H}}$$

$$= \sum_{i=1}^{k}\sum_{j=1}^{\infty}\langle\left(\langle\Psi, e_i\rangle_{\mathcal{H}}Y - [\langle\Psi, e_i\rangle_{\mathcal{H}}, Y]\right)e_i, \langle\phi, e_j\rangle_{\mathcal{H}}e_j\rangle_{\mathcal{H}}$$

$$= \sum_{i=1}^{k}\left(\langle\Psi, e_i\rangle_{\mathcal{H}}Y - [\langle\Psi, e_i\rangle_{\mathcal{H}}, Y]\right)\langle\phi, e_i\rangle_{\mathcal{H}},$$

where we recall that each process $\langle\Psi, e_i\rangle_{\mathcal{H}}Y - [\langle\Psi, e_i\rangle_{\mathcal{H}}, Y]$ is a martingale, hence too is

$\left(\langle\Psi, e_i\rangle_{\mathcal{H}}Y - [\langle\Psi, e_i\rangle_{\mathcal{H}}, Y]\right)\langle\phi, e_i\rangle_{\mathcal{H}}$, and therefore the finite sum is a martingale.

To comment on the uniqueness of $[\mathcal{P}_k \Psi, Y]$, we invoke Lemma 2.3. Suppose that Π is an $\mathcal{H}$-valued process which is continuous, adapted, and of bounded-variation satisfying $\Pi_0 = 0 \ \mathbb{P} - a.s.$. Moreover, suppose that

$$(\mathcal{P}_k \Psi)Y - \Pi$$

is an $\mathcal{H}$-valued martingale. Take any basis vector e_j. Then

$$\langle (\mathcal{P}_k \Psi)Y - \Pi, e_j \rangle_{\mathcal{H}}$$

is a real valued martingale, but this is just

$$\sum_{i=1}^{k} \langle \Psi, e_i \rangle_{\mathcal{H}} \langle e_i, e_j \rangle_{\mathcal{H}} Y - \langle \Pi, e_j \rangle_{\mathcal{H}}. \qquad (2.33)$$

From the regularity of Π, we have again that $\langle \Pi, e_j \rangle_{\mathcal{H}}$ is continuous, adapted, and of bounded-variation satisfying $\langle \Pi_0, e_j \rangle_{\mathcal{H}} = 0 \ \mathbb{P} - a.s.$. There are two cases here: $j \leq k$ and $j > k$. In the latter case we have that the first term in (2.33) is null so in particular $\langle \Pi, e_j \rangle_{\mathcal{H}}$ is a martingale. Thus $\langle \Pi, e_j \rangle_{\mathcal{H}} \in \mathcal{M}_c$ and is of bounded-variation, so by Lemma A.1 it is $\mathbb{P} - a.s.$ constant. As the process starts from zero, then it is indistinguishable from the null process. Therefore $\langle \Pi, e_j \rangle_{\mathcal{H}}$ is indistinguishable from $\langle [\mathcal{P}_k \Psi, Y], e_j \rangle$, which is similarly zero (recall Definition 2.17).

In the alternative case $j \leq k$, the real valued martingale (2.33) is given by

$$\langle \Psi, e_j \rangle_{\mathcal{H}} Y - \langle \Pi, e_j \rangle_{\mathcal{H}}.$$

Therefore $\langle \Pi, e_j \rangle_{\mathcal{H}}$ satisfies the requirements of the cross-variation process $[\langle \Psi, e_j \rangle_{\mathcal{H}}, Y]$ and hence is indistinguishable from it. This, however, is simply $\langle [\mathcal{P}_k \Psi, Y], e_k \rangle$ from its definition. Combining with Lemma 2.3, the theorem is proven. $\qquad \square$

It remains for us to define the cross-variation process $[\Psi, Y]$. Referencing Lemma A.2, we know that $\mathbb{P} - a.s.$ for any $t \geq 0$,

$$[\langle \Psi, e_i \rangle_{\mathcal{H}}, Y]_t^2 \leq [\langle \Psi, e_i \rangle_{\mathcal{H}}]_t [Y]_t,$$

hence, using the nondecreasing property of quadratic variation,

$$\sup_{t \in [0,T]} [\langle \Psi, e_i \rangle_{\mathcal{H}}, Y]_t^2 \leq \sup_{t \in [0,T]} \left([\langle \Psi, e_i \rangle_{\mathcal{H}}]_t [Y]_t \right)$$

$$\leq \left(\sup_{t \in [0,T]} [\langle \Psi, e_i \rangle_{\mathcal{H}}]_t \right) \left(\sup_{t \in [0,T]} [Y]_t \right)$$

$$= [\langle \Psi, e_i \rangle_{\mathcal{H}}]_T [Y]_T.$$

Moreover for $j < k$,

$$\sup_{t \in [0,T]} \left\| [\mathcal{P}_k \Psi, Y]_t - [\mathcal{P}_j \Psi, Y]_t \right\|_{\mathcal{H}}^2 = \sup_{t \in [0,T]} \left\| \sum_{i=j+1}^{k} [\langle \Psi, e_i \rangle_{\mathcal{H}}, Y]_t e_i \right\|_{\mathcal{H}}^2$$

$$= \sup_{t \in [0,T]} \sum_{i=j+1}^{k} [\langle \Psi, e_i \rangle_{\mathcal{H}}, Y]_t^2$$

$$\leq \sum_{i=j+1}^{k} \sup_{t \in [0,T]} [\langle \Psi, e_i \rangle_{\mathcal{H}}, Y]_t^2$$

$$\leq \sum_{i=j+1}^{k} [\langle \Psi, e_i \rangle_{\mathcal{H}}]_T [Y]_T$$

$$= [\mathcal{P}_k \Psi - \mathcal{P}_j \Psi]_T [Y]_T$$

using (2.30) and (2.31) for the last line. We look to follow a similar approach to Proposition 2.12, taking the expectation and showing a Cauchy property. It is unknown if this resulting term is integrable, so in order to take the expectation we introduce the localizing times

$$\tau_n := n \wedge \inf \{ 0 \leq t < \infty : [Y]_t \geq n \}.$$

Then

$$\mathbb{E} \left(\left(\sup_{t \in [0,T]} \left\| [\mathcal{P}_k \Psi, Y]_t \mathbb{1}_{t \leq \tau_n} - [\mathcal{P}_j \Psi, Y]_t \mathbb{1}_{t \leq \tau_n} \right\|_{\mathcal{H}} \right)^2 \right)$$

$$= \mathbb{E} \left(\sup_{t \in [0,T]} \left\| [\mathcal{P}_k \Psi, Y]_t \mathbb{1}_{t \leq \tau_n} - [\mathcal{P}_j \Psi, Y]_t \mathbb{1}_{t \leq \tau_n} \right\|_{\mathcal{H}}^2 \right)$$

$$\leq \mathbb{E} \left([\mathcal{P}_k \Psi - \mathcal{P}_j \Psi]_T [Y]_T \mathbb{1}_{t \leq \tau_n} \right)$$

$$\leq n \mathbb{E} \left([\mathcal{P}_k \Psi - \mathcal{P}_j \Psi]_T \right),$$

which was shown to approach zero as $j \to \infty$, uniformly in k, in Proposition 2.12. We have thus demonstrated that for every $n \in \mathbb{N}$ and $T \geq 0$, the sequence $([\mathcal{P}_k \Psi, Y].\mathbb{1}_{.\leq \tau_n})$ is Cauchy in the Banach Space $L^2(\Omega; L^\infty([0, T]; \mathcal{H}))$. We can therefore extract a subsequence which converges $\mathbb{P} - a.s.$ in $L^\infty([0, T]; \mathcal{H})$. To remove the truncation we introduce the sets

$$A_n := \{\omega \in \Omega : \tau_n(\omega) \geq T\}.$$

Then on every A_n, there exists a subsequence of $([\mathcal{P}_k \Psi, Y])$ which is $\mathbb{P} - a.s.$ (within A_n) convergent in $L^\infty([0, T]; \mathcal{H})$. It should be noted that the choice of subsequence may be dependent on n, hence the separation, and also that this convergence is now of continuous processes so can be taken in $C([0, T]; \mathcal{H})$. We look to upgrade this convergence to be of the whole sequence, by another Cauchy argument. Fix any $\varepsilon > 0$, and we look to show the existence of a $J \in \mathbb{N}$ such that for all $k \geq J$,

$$\sup_{t \in [0,T]} \|[\mathcal{P}_k \Psi, Y]_t - [\mathcal{P}_J \Psi, Y]_t\|_{\mathcal{H}}^2 < \varepsilon$$

or, equivalently, as already shown in the proof,

$$\sup_{t \in [0,T]} \sum_{i=J+1}^{k} \left[\langle \Psi, e_i \rangle_{\mathcal{H}}, Y \right]_t^2 < \varepsilon.$$

Let the convergent subsequence be indexed by k_m. This convergence implies that the subsequence is Cauchy, so there exists a J_m such that for all $k_m > J_m$,

$$\sup_{t \in [0,T]} \|[\mathcal{P}_{k_m} \Psi, Y]_t - [\mathcal{P}_{J_m} \Psi, Y]_t\|_{\mathcal{H}}^2 < \varepsilon$$

or equivalently

$$\sup_{t \in [0,T]} \sum_{i=J_m+1}^{k_m} \left[\langle \Psi, e_i \rangle_{\mathcal{H}}, Y \right]_t^2 < \varepsilon.$$

We now set $J := J_m$ and argue that for every $k > J$ there exists a $k_m > k$ such that

$$\sup_{t \in [0,T]} \sum_{i=J+1}^{k} \left[\langle \Psi, e_i \rangle_{\mathcal{H}}, Y \right]_t^2 \leq \sup_{t \in [0,T]} \sum_{i=J+1}^{k_m} \left[\langle \Psi, e_i \rangle_{\mathcal{H}}, Y \right]_t^2 < \varepsilon$$

which proves the Cauchy property. Hence on each A_n, the sequence $([\mathcal{P}_k \Psi, Y])$ is $\mathbb{P} - a.s.$ convergent in $C([0, T]; \mathcal{H})$. We use that

$$\mathbb{P}\left(\bigcup_n A_n \right) = 1 - \mathbb{P}\left(\bigcap_n A_n^c \right) = 1 - \mathbb{P}\left(\bigcap_n \{\omega \in \Omega : \tau_n(\omega) < T\} \right)$$

$$= 1 - \lim_{n \to \infty} \mathbb{P}\left(\{\omega \in \Omega : \tau_n(\omega) < T\} \right)$$

owing to the fact that the (τ_n) are $\mathbb{P}-a.s.$ increasing. Moreover, this limit is zero given that the (τ_n) approach infinity. Thus we obtain the $\mathbb{P}-a.s.$ convergence on Ω.

Definition 2.18 For $\Psi \in \mathcal{M}_c^2(\mathcal{H})$ and $Y \in \mathcal{M}_c^2$, we define the cross-variation process $[\Psi, Y]$ by

$$[\Psi, Y] := \sum_{i=1}^{\infty} [\langle \Psi, e_i \rangle_{\mathcal{H}}, Y] e_i$$

$\mathbb{P}-a.s.$ for the limit taken in $C([0, T]; \mathcal{H})$.

Our next question is then very natural: Do we have the corresponding characterization as in Proposition 2.13? The fact that this cross-variation is continuous, adapted, and starting from zero is proven as in Proposition 2.12. In fact the martingality of

$$\Psi Y - [\Psi, Y] \tag{2.34}$$

is again proven near identically, as we use that

$$(\mathcal{P}_k \Psi) Y - \sum_{i=1}^{k} [\langle \Psi, e_i \rangle_{\mathcal{H}}, Y] e_i$$

is a martingale, and the convergence of this at each $t \geq 0$ in $L^1(\Omega; \mathcal{H})$ to (2.34). It is the bounded-variation which proves problematic. Possibly the most logical approach is to show that the finite sums are of uniformly bounded total variation, through an argument like

$$V_{\mathcal{H}}^T([\mathcal{P}_k \Psi, Y]) = V_{\mathcal{H}}^T \left(\sum_{i=1}^{k} [\langle \Psi, e_i \rangle_{\mathcal{H}}, Y] e_i \right) \leq \sum_{i=1}^{k} V_{\mathcal{H}}^T \left([\langle \Psi, e_i \rangle_{\mathcal{H}}, Y] e_i \right)$$

$$= \sum_{i=1}^{k} V_{\mathbb{R}}^T \left([\langle \Psi, e_i \rangle_{\mathcal{H}}, Y] \right).$$

Then from Lemma A.3, we have that

$$V_{\mathbb{R}}^T \left([\langle \Psi, e_i \rangle_{\mathcal{H}}, Y] \right) \leq \frac{1}{2} \left([\langle \Psi, e_i \rangle_{\mathcal{H}}]_T + [Y]_T \right),$$

but this means taking $\sum_{i=1}^{k} [Y]_T$, which will explode as $k \to \infty$. In lieu of this bounded-variation, we do of course have the weaker property that every projection is of bounded-variation. In fact, this is enough for uniqueness.

Suppose that Π is an $\mathcal{H}$-valued process which is continuous, adapted, satisfying $\Pi_0 = 0$ and such that for every basis vector e_j, $\langle \Pi, e_j \rangle$ is of bounded-variation $\mathbb{P} - a.s.$. Moreover suppose that

$$\Psi Y - \Pi$$

is an $\mathcal{H}$-valued martingale. Take any basis vector e_j. Then

$$\langle \Psi Y - \Pi, e_j \rangle_{\mathcal{H}}$$

is a real valued martingale, but this is just

$$\sum_{i=1}^{\infty} \langle \Psi, e_i \rangle_{\mathcal{H}} \langle e_i, e_j \rangle_{\mathcal{H}} Y - \langle \Pi, e_j \rangle_{\mathcal{H}}$$

or simply

$$\langle \Psi, e_j \rangle_{\mathcal{H}} Y - \langle \Pi, e_j \rangle_{\mathcal{H}}.$$

Thus $\langle \Pi, e_j \rangle_{\mathcal{H}}$ is indistinguishable from $[\langle \Psi, e_j \rangle_{\mathcal{H}}, Y]$ which is simply $\langle [\Psi, Y], e_k \rangle$ from its definition. Combining with Lemma 2.3, we deduce the uniqueness in this class, so have proven the following:

Proposition 2.14 *For $\Psi \in \mathcal{M}_c^2(\mathcal{H})$ and $Y \in \mathcal{M}_c^2$, $[\Psi, Y]$ is the unique, continuous, adapted $\mathcal{H}$-valued process satisfying $[\Psi, Y]_0 = 0$ $\mathbb{P} - a.s.$ such that for every basis vector e_j, $\langle [\Psi, Y], e_j \rangle_{\mathcal{H}}$ is of bounded-variation $\mathbb{P} - a.s.$ and*

$$\Psi Y - [\Psi, Y]$$

is an $\mathcal{H}$-valued martingale.

We now state and prove the analogous result to Lemma 2.5, which will be necessary in the Itô–Stratonovich conversion.

Lemma 2.6 *Suppose that (Ψ^n) is a sequence of martingales in $\mathcal{M}_c^2(\mathcal{H})$ which at every time $t \geq 0$ converges in $L^2(\Omega; \mathcal{H})$ to some Ψ_t. Let $Y \in \mathcal{M}_c^2$. Suppose in addition that at any time $t \geq 0$, the sequence $([\Psi^n, Y]_t)$ converges to some L_t in $L^1(\Omega; \mathbb{R})$, where L is a continuous, adapted process, and for every basis vector e_j, $\langle L, e_j \rangle_{\mathcal{H}}$ is of bounded-variation $\mathbb{P} - a.s.$. Then $\Psi \in \mathcal{M}_c^2(\mathcal{H})$ and $[\Psi, Y]$ is indistinguishable from L.*

Proof For each n the martingale in question is the $\mathcal{H}$-valued one

$$\Psi^n Y - [\Psi^n, Y].$$

We use again that the $L^1(\Omega; \mathcal{H})$ limit preserves martingality and that by Hölder's inequality the $L^1(\Omega; \mathcal{H})$ limit of $\Psi_t^n Y_t$ is $\Psi_t Y_t$. With the same arguments as in Lemma 2.5 we conclude the proof. □

If the given processes were only (continuous) local martingales, then we can make a slightly modified version of the definition. Assuming without loss of generality that Ψ and Y are locally square integrable (see the discussion after Definition 2.9), localized by stopping times (R_n) and (T_n), respectively, then for a new sequence of stopping times defined by $\tau_n = R_n \wedge T_n$, the stopped processes Ψ^{τ_n} and Y^{τ_n} are genuine square integrable martingales (in their respective spaces), so the cross-variation $[\Psi^{\tau_n}, Y^{\tau_n}]$ can be defined. The canonical localization procedure is evident once more, as the (τ_n) tends to infinity almost surely, the consistency conditions that for $m \le n$ and $t \le \tau_m$ we have

$$\Psi_t^{\tau_m} = \Psi_t^{\tau_n} \qquad \text{and} \qquad Y_t^{\tau_m} = Y_t^{\tau_n}$$

allow us once more to define the process at any t by

$$[\Psi, Y]_t := \lim_{n \to \infty} [\Psi^{\tau_n}, Y^{\tau_n}]_t \tag{2.35}$$

for the limit taken $\mathbb{P} - a.s.$ in $\mathcal{H}$. Then the process

$$\Psi Y - [\Psi, Y]$$

is itself a local martingale, localized by the stopping times (τ_n). The argument justifying this is identical to Proposition 2.9, from which it is similarly clear that

$$[\Psi, Y]^{\tau_n} = [\Psi^{\tau_n}, Y^{\tau_n}].$$

In the traditional way, these notions can all be extended to semimartingales (that is, a martingale plus a bounded-variation process). The quadratic and cross-variations of such semi-martingales are then simply the quadratic/cross-variation of the corresponding martingale parts. To this end we introduce the notations $\bar{\mathcal{M}}_c^2$ and $\bar{\mathcal{M}}_c^2(\mathcal{H})$ to be the corresponding spaces of square integrable continuous semimartingales, and similarly $\bar{\mathcal{M}}_c$, $\bar{\mathcal{M}}_c(\mathcal{H})$ to be the spaces of continuous semimartingales.

2.5 Integration with Respect to Cylindrical Brownian Motion

For our analysis now we will need to make reference to two distinct Hilbert spaces: one over which $\mathcal{W}$ is a cylindrical Brownian motion, and the other in which our integrand maps to. Henceforth we introduce $\mathfrak{U}$ as the Hilbert space over which $\mathcal{W}$

is a cylindrical Brownian motion. We shall take (e_i) as an orthonormal basis over $\mathfrak{U}$ and (a_i) an orthonormal basis over $\mathcal{H}$.

Definition 2.19 Denote by $I_T^{\mathcal{H}}(\mathcal{W})$ the class of progressively measurable operator valued processes B belonging to the set $L^2(\Omega \times [0, T]; \mathscr{L}^2(\mathfrak{U}; \mathcal{H}))$. Measurability here is again defined with respect to the Borel sigma algebra on $\mathscr{L}^2(\mathfrak{U}; \mathcal{H})$. The class of processes B such that $B \in I_T^{\mathcal{H}}(\mathcal{W})$ for all T will be denoted by $I^{\mathcal{H}}(\mathcal{W})$.

Note that we make no explicit reference to $\mathfrak{U}$, the space on which $\mathcal{W}$ is a cylindrical Brownian motion. This is because, in practice, the space $\mathfrak{U}$ will be arbitrarily chosen; this shall be discussed later. Recall (2.22) that if $\mathcal{W}$ is a cylindrical Brownian motion over $\mathfrak{U}$, it can be formally represented by

$$\mathcal{W}(t) = \sum_{i=1}^{\infty} e_i W_t^i, \tag{2.36}$$

where the (W^i) are standard independent one dimensional Brownian motions.

Definition 2.20 For $B \in I^{\mathcal{H}}(\mathcal{W})$, we define the Itô stochastic integral

$$\int_0^t B(s)d\mathcal{W}_s \tag{2.37}$$

as the $\mathcal{H}$-valued random variable

$$\sum_{i=1}^{\infty} \int_0^t B_{e_i}(s)dW_s^i, \tag{2.38}$$

where each integral is defined as in Definition 2.6 and the infinite sum is taken in $L^2(\Omega; \mathcal{H})$.

The immediate response to this definition is to prove that (2.38) is well defined; that is, the integrals are well defined, as is the limit. First for each i, B_{e_i} is trivially in $L^2(\Omega \times [0, T]; \mathcal{H})$ as this norm is bounded by the $L^2(\Omega \times [0, T]; \mathscr{L}^2(\mathfrak{U}; \mathcal{H}))$ norm of B. The progressive measurability is inherited from that of B.

In order to show that the limit of partial sums is well defined, we proceed similarly to the method applied for (2.19) and argue that the sequence of partial sums is Cauchy. Observe that

$$\left\| \sum_{i=m}^{n} \int_0^t B_{e_i}(s)dW_s^i \right\|_{L^2(\Omega;\mathcal{H})}^2 = \mathbb{E} \left\| \sum_{i=m}^{n} \int_0^t B_{e_i}(s)dW_s^i \right\|_{\mathcal{H}}^2$$

$$= \sum_{i=m}^{n} \mathbb{E} \int_0^t \left\| B_{e_i}(s) \right\|_{\mathcal{H}}^2 ds$$

having applied the Itô isometry Proposition 2.3 to the above. But by the assumption that $B \in L^2\big(\Omega \times [0, t]; \mathscr{L}^2(\mathfrak{U}; \mathcal{H})\big)$, we know

$$\mathbb{E} \int_0^t \sum_{i=1}^\infty \big\| B_{e_i}(s) \big\|_{\mathcal{H}}^2 ds < \infty$$

and thus, by Tonelli's theorem regarding the infinite sum as an integral with respect to the counting measure,

$$\sum_{i=1}^\infty \mathbb{E} \int_0^t \big\| B_{e_i}(s) \big\|_{\mathcal{H}}^2 ds < \infty$$

demonstrating that the sequence of partial sums is indeed Cauchy in $L^2(\Omega; \mathcal{H})$. Of course the $L^2(\Omega; \mathcal{H})$ norm of the limit is the limit of the $L^2(\Omega; \mathcal{H})$ norms, so we have justified the following.

Proposition 2.15 *For $B \in I^{\mathcal{H}}(\mathcal{W})$, we have*

$$\mathbb{E} \left(\left\| \int_0^t B(s) d\mathcal{W}_s \right\|_{\mathcal{H}}^2 \right) = \mathbb{E} \left(\int_0^t \| B(s) \|_{\mathscr{L}^2(\mathfrak{U}; \mathcal{H})}^2 ds \right).$$

It is worth noting that while we impose the condition

$$\mathbb{E} \left(\int_0^t \sum_{i=1}^\infty \big\| B_{e_i}(s) \big\|_{\mathcal{H}}^2 ds \right) < \infty,$$

one may instead require the weaker condition

$$\int_0^t \sum_{i=1}^\infty \big\| B_{e_i}(s) \big\|_{\mathcal{H}}^2 ds < \infty \qquad \mathbb{P} - a.s. \tag{2.39}$$

or equivalently that $B : \Omega \to L^2\big([0, t]; \mathscr{L}^2(\mathfrak{U}; \mathcal{H})\big)$ for $\mathbb{P} - a.e.$ ω. Our formulation follows the classical construction as laid out in Sect. 2.1, ensuring that the integral is a genuine square integrable martingale, which one anticipates would be lost with just the assumption (2.39). We can just as straightforwardly follow the arguments from Definition 2.9, which are laid out here. The extension is pertinent in applications, and this brief will closely monitor the necessity of localization in the abstract frameworks of Chaps. 3 and 4.

Definition 2.21 Denote by $\overline{I}_T^{\mathcal{H}}(\mathcal{W})$ the class of progressively measurable operator valued processes B such that $B(\omega)$ belongs to the set $L^2\big([0, T]; \mathscr{L}^2(\mathfrak{U}; \mathcal{H})\big)$ for $\mathbb{P} - a.e.$ ω. The class of processes B such that $B \in \overline{I}_T^{\mathcal{H}}(\mathcal{W})$ for all T will be denoted by $\overline{I}^{\mathcal{H}}(\mathcal{W})$.

Using the template for Definition 2.10, for a process $B \in \overline{I}^{\mathcal{H}}(\mathcal{W})$, let us introduce

$$\tau_n := n \wedge \inf\{0 \leq t < \infty : \int_0^t \|B(s)\|^2_{\mathscr{L}^2(\mathfrak{U};\mathcal{H})} ds \geq n\}$$

taking the convention that the infimum of the empty set is infinite. The (τ_n) are stopping times as they are simply first hitting times of the continuous and adapted random variable $\int_0^t \|B(s)\|^2_{\mathscr{L}^2(\mathfrak{U};\mathcal{H})} ds$. These times tend to infinity $\mathbb{P} - a.s.$ by condition (2.39). Now define the truncated processes B^n as

$$B^n(t) := B(t)\mathbb{1}_{t \leq \tau_n},$$

and using the fact that for $m \leq n$, and $t \leq \tau_m$, we have

$$B(t)\mathbb{1}_{t \leq \tau_n} = B(t)\mathbb{1}_{t \leq \tau_m}$$

we can make the following consistent definition.

Definition 2.22 In the setting described, we define

$$\int_0^t B(s)d\mathcal{W}_s := \lim_{n \to \infty} \int_0^t B^n(s)d\mathcal{W}_s \qquad (2.40)$$

$\mathbb{P} - a.s.$ ω in $\mathcal{H}$.

We will have no reservations in writing (2.40) as a formal expression

$$\sum_{i=1}^{\infty} \int_0^t B_{e_i}(s)d\mathbb{W}_s^i. \qquad (2.41)$$

We say the expression is only formal, as the infinite sum in (2.41) is *not* the $L^2(\Omega; \mathcal{H})$ limit of the partial sums of the local martingales as presented. We understand (2.41) only by (2.40), that is, by the limit as $n \to \infty$ of the infinite sum in $L^2(\Omega; \mathcal{H})$ of the genuine square integrable martingales given by stopping the local martingales at τ_n.

We now look to show a series of properties of this integral which were shown for a one dimensional Brownian motion across the earlier sections. The first is the corresponding result of Theorem 2.2.

Proposition 2.16 *Suppose that $\mathcal{H}_1, \mathcal{H}_2$ are Hilbert spaces such that $B \in \overline{I}^{\mathcal{H}_1}(\mathcal{W})$ and $T \in \mathscr{L}(\mathcal{H}_1; \mathcal{H}_2)$. Then the process TB defined by*

$$TB_{e_i}(s, \omega) = T\big(B_{e_i}(s, \omega)\big)$$

belongs to $\overline{I}^{\mathcal{H}_2}(\mathcal{W})$. In addition, we have that

$$T\left(\int_0^t B(s)d\mathcal{W}_s\right) = \int_0^t T B(s)d\mathcal{W}_s.$$

Moreover, the two integrals are defined as limits $\mathbb{P} - a.s.$ *with respect to the same stopping times, and if* $B \in \mathcal{I}^{\mathcal{H}_1}(\mathcal{W})$, *then* $T B \in \mathcal{I}^{\mathcal{H}_2}(\mathcal{W})$.

Proof Assume at first that $B \in \overline{\mathcal{I}}^{\mathcal{H}_1}(\mathcal{W})$. We shall prove first that $T B \in \overline{\mathcal{I}}^{\mathcal{H}_2}(\mathcal{W})$. The progressive measurability is preserved under the continuity of T, and letting C be such that for any $\phi \in \mathcal{H}_1$,

$$\|T\phi\|_{\mathcal{H}_2}^2 \le C\|\phi\|_{\mathcal{H}_1}^2,$$

which exists owing to the boundedness of T, we have

$$\int_0^t \sum_{i=1}^\infty \|T B_{e_i}(s)\|_{\mathcal{H}_2}^2 ds \le C \int_0^t \sum_{i=1}^\infty \|B_{e_i}(s)\|_{\mathcal{H}_1}^2 ds < \infty \qquad (2.42)$$

holding $\mathbb{P} - a.s.$ as $B \in \overline{\mathcal{I}}^{\mathcal{H}_1}(\mathcal{W})$. In addition for any stopping time τ_n as in Definition 2.22,

$$\mathbb{E}\left(\int_0^t \sum_{i=1}^\infty \|(T B_{e_i}(s))\mathbb{1}_{s\le\tau_n}\|_{\mathcal{H}_2}^2 ds\right) = \mathbb{E}\left(\int_0^t \sum_{i=1}^\infty \|T(B_{e_i}(s)\mathbb{1}_{s\le\tau_n})\|_{\mathcal{H}_2}^2 ds\right)$$

$$\le C\mathbb{E}\left(\int_0^t \sum_{i=1}^\infty \|B_{e_i}(s)\mathbb{1}_{s\le\tau_n}\|_{\mathcal{H}_1}^2 ds\right)$$

$$< \infty.$$

So the new stochastic integral

$$\int_0^t T B(s)d\mathcal{W}_s$$

can be constructed using the same sequence of stopping times. We will freely use the linearity of T to commute it with the indicator function. To carry T through the integral, it is sufficient to show that for B^n as in Definition 2.22, then

$$T\left(\sum_{i=1}^\infty \int_0^t B_{e_i}^n(s)d\mathcal{W}_s^i\right) = \sum_{i=1}^\infty \int_0^t T B_{e_i}^n(s)d\mathcal{W}_s^i, \qquad (2.43)$$

where the left hand side limit is taken in $L^2(\Omega; \mathcal{H}_1)$ and the right side in $L^2(\Omega; \mathcal{H}_2)$. From the $L^2(\Omega; \mathcal{H}_1)$ limit there exists a subsequence convergent $\mathbb{P} - a.e.$ in $\mathcal{H}_1$. Working with this subsequence, we can pass the $\mathbb{P} - a.s.$ limit through the continuous T such that it is now the $\mathbb{P} - a.s.$ limit in $\mathcal{H}_2$. Linearity of T allows us to pass through the summation as well, so we have now that

$$T\left(\sum_{i=1}^{\infty}\int_0^t B_{e_i}^n(s)dW_s^i\right) = \lim_{n_k\to\infty}\sum_{i=1}^{n_k}T\left(\int_0^t B_{e_i}^n(s)dW_s^i\right) \tag{2.44}$$

for the limit $\mathbb{P} - a.e.$ of the subsequence indexed by (n_k). Applying Theorem 2.2, we can commute T with the integral on the right side of (2.44). However we have justified already that the limit over the whole sequence in $L^2(\Omega; \mathcal{H}_2)$ exists, agreeing with the $L^2(\Omega; \mathcal{H}_2)$ limit of the subsequence, which in turn agrees with the $\mathbb{P} - a.s.$ limit. Thus (2.43) is verified, completing the proof. In the case where $B \in \mathcal{I}^{\mathcal{H}_1}(\mathcal{W})$, it is clear from (2.42) that $TB \in \mathcal{I}^{\mathcal{H}_2}(\mathcal{W})$. □

Proposition 2.17 *Let $B \in \bar{\mathcal{I}}^{\mathcal{H}}(\mathcal{W})$ and $\phi : \Omega \to \mathcal{H}$ be $\mathcal{F}_0$-measurable. Then for every $t > 0$ we have that*

$$\left\langle\int_0^t B_r dW_r, \phi\right\rangle_{\mathcal{H}} = \int_0^t \langle B_r, \phi\rangle_{\mathcal{H}} dW_r \tag{2.45}$$

$\mathbb{P} - a.s.$. *Moreover if $\eta : \Omega \to \mathbb{R}$ is $\mathcal{F}_0$-measurable, then $\eta B \in \bar{\mathcal{I}}^{\mathcal{H}}(\mathcal{W})$, and for every $t > 0$ we have that*

$$\eta\int_0^t B_r dW_r = \int_0^t \eta B_r dW_r \tag{2.46}$$

$\mathbb{P} - a.s.$.

Proof First we make clear that $\langle B, \phi\rangle_{\mathcal{H}}$ is understood as a process defined by the mapping

$$(e_i, \omega, t) \to \langle B_t(e_i, \omega), \phi(\omega)\rangle_{\mathcal{H}}.$$

The fact that $\langle B, \phi\rangle_{\mathcal{H}} \in \bar{\mathcal{I}}^{\mathbb{R}}(\mathcal{W})$ is completely analogous to Proposition 2.4, where the progressive measurability follows as the mapping

$$\langle B_\cdot, \phi\rangle_{\mathcal{H}} : (t, \omega, \tilde{\omega}) \to \langle B_t(\omega), \phi(\tilde{\omega})\rangle_{\mathcal{H}}$$

is $\mathcal{B}([0, T]) \times \mathcal{F}_T \times \mathcal{F}_0$ measurable as a mapping into $\mathscr{L}^2(\mathfrak{U}; \mathbb{R})$. Similarly we have that

$$\left\|\langle B_t(\omega), \phi(\omega)\rangle_{\mathcal{H}}\right\|_{\mathscr{L}^2(\mathfrak{U};\mathbb{R})} \le \|\phi(\omega)\|_{\mathcal{H}}\|B_t(\omega)\|_{\mathscr{L}^2(\mathfrak{U};\mathcal{H})},$$

which is sufficient to justify that $\langle B, \phi\rangle_{\mathcal{H}} \in \bar{\mathcal{I}}^{\mathbb{R}}(\mathcal{W})$. The extension of Proposition 2.5 to this result is then identical to the extension of Theorem 2.2 to 2.16, so we conclude the proof of (2.45) here. The property (2.46) follows identically in analogy with Proposition 2.6. We make explicit that ηB is defined by the mapping

$$e_i \times \omega \times t \to \eta(\omega)B_t(e_i, \omega).$$

$\square$

Remark 2.4 Although we have not explicitly addressed the construction of the integral over a time interval $[s, t]$ where $s > 0$, this can be done without any extra difficulty just as in the standard real valued case. If we were to just consider the integral over $[s, t]$ in Proposition 2.17, then the results extend to any $\mathcal{F}_s$-measurable ϕ, η in Proposition 2.17. To show this we revisit Proposition 2.4, appreciating that the $\mathcal{F}_s$-measurability does not disturb the measurability requirements of the simple process.

We also extend the Stochastic Dominated Convergence Theorem to this setting.

Lemma 2.7 Let (B^n) be a sequence in $\bar{I}^{\mathcal{H}}(\mathcal{W})$ such that there exist processes $B : \Omega \times [0, \infty) \to \mathscr{L}^2(\mathfrak{U}; \mathcal{H})$ and $Q \in L^2([0, \infty); \mathbb{R})$ $\mathbb{P} - a.s.$, with the properties that for every $T > 0$, $\mathbb{P} \times \lambda - a.e.$ $(\omega, t) \in \Omega \times [0, T]$:

1. $\left\| B_t^n(\omega) \right\|_{\mathscr{L}^2(\mathfrak{U};\mathcal{H})} \leq |Q_t(\omega)|$ for all $n \in \mathbb{N}$.
2. $(B_t^n(\omega))$ is convergent to $B_t(\omega)$ in $\mathscr{L}^2(\mathfrak{U}; \mathcal{H})$.

Then $B \in \bar{I}^{\mathcal{H}}(\mathcal{W})$, and for every $t > 0$, there exists a subsequence indexed by (n_k) such that

$$\lim_{n_k \to \infty} \int_0^t B_r^{n_k} d\mathcal{W}_r = \int_0^t B_r d\mathcal{W}_r \tag{2.47}$$

$\mathbb{P} - a.s.$.

Proof The proof is identical to that of Lemma 2.1, simply replacing $\mathcal{H}$ by $\mathscr{L}^2(\mathfrak{U}; \mathcal{H})$ when dealing with the integrands and using the appropriate Itô isometry Proposition 2.15. $\square$

We now shift attentions to results regarding the martingale properties of the integral.

Proposition 2.18 For $B \in I^{\mathcal{H}}(\mathcal{W})$, the Itô stochastic integral

$$\int_0^t B(s)d\mathcal{W}_s$$

belongs to $\mathcal{M}_c^2(\mathcal{H})$. If $B \in I^{\mathcal{H}}(\mathcal{W})$ only, then the integral is a continuous local martingale.

Proof This follows immediately from Propositions 2.7, 2.8 and, subsequently, Proposition 2.9. $\square$

The martingality of the integral also allows us to consider the quadratic variation as defined in Definition 2.16.

Proposition 2.19 *For $B \in I^{\mathcal{H}}(\mathcal{W})$, we have that*

$$\left[\int_0^{\cdot} B_r d\mathcal{W}_r\right]_t = \int_0^t \|B_r\|_{\mathscr{L}^2(\mathfrak{U};\mathcal{H})}^2 dr. \tag{2.48}$$

Proof At each time t, the integral

$$\int_0^t B_r d\mathcal{W}_r$$

is defined to be the $L^2(\Omega; \mathcal{H})$ limit of the sequence

$$\sum_{i=1}^{n} \int_0^t B_r(e_i) dW_r^i.$$

We look to infer the quadratic variation of this sequence of processes using Proposition 2.10, to then apply Lemma 2.5. We claim that

$$\left[\sum_{i=1}^{n} \int_0^{\cdot} B_r(e_i) dW_r^i\right]_t = \int_0^t \sum_{i=1}^{n} \|B_r(e_i)\|_{\mathcal{H}}^2 dr,$$

which is to say

$$\left\|\sum_{i=1}^{n} \int_0^t B_r(e_i) dW_r^i\right\|_{\mathcal{H}}^2 - \int_0^t \sum_{i=1}^{n} \|B_r(e_i)\|_{\mathcal{H}}^2 dr \tag{2.49}$$

is a martingale. For the orthonormal basis (a_k) of $\mathcal{H}$,

$$\left\|\sum_{i=1}^{n} \int_0^t B_r(e_i) dW_r^i\right\|_{\mathcal{H}}^2 - \int_0^t \sum_{i=1}^{n} \|B_r(e_i)\|_{\mathcal{H}}^2 dr$$

$$= \sum_{k=1}^{\infty} \left\langle \sum_{i=1}^{n} \int_0^t B_r(e_i) dW_r^i, a_k \right\rangle_{\mathcal{H}}^2 - \int_0^t \sum_{i=1}^{n} \|B_r(e_i)\|_{\mathcal{H}}^2 dr$$

$$= \sum_{k=1}^{\infty} \sum_{i=1}^{n} \sum_{j=1}^{n} \left\langle \int_0^t B_r(e_i) dW_r^i, a_k \right\rangle_{\mathcal{H}} \left\langle \int_0^t B_r(e_j) dW_r^j, a_k \right\rangle_{\mathcal{H}}$$

$$- \int_0^t \sum_{i=1}^{n} \|B_r(e_i)\|_{\mathcal{H}}^2 dr$$

$$= \left(\sum_{k=1}^{\infty} \sum_{i=1}^{n} \left\langle \int_0^t B_r(e_i) dW_r^i, a_k \right\rangle_{\mathcal{H}}^2 - \int_0^t \sum_{i=1}^{n} \| B_r(e_i) \|_{\mathcal{H}}^2 dr \right)$$

$$+ \left(\sum_{k=1}^{\infty} \sum_{i \neq j} \left\langle \int_0^t B_r(e_i) dW_r^i, a_k \right\rangle_{\mathcal{H}} \left\langle \int_0^t B_r(e_j) dW_r^j, a_k \right\rangle_{\mathcal{H}} \right)$$

$$= \sum_{i=1}^{n} \left(\left\| \int_0^t B_r(e_i) dW_r^i \right\|_{\mathcal{H}}^2 - \int_0^t \| B_r(e_i) \|_{\mathcal{H}}^2 dr \right)$$

$$+ \left(\sum_{k=1}^{\infty} \sum_{i \neq j} \left\langle \int_0^t B_r(e_i) dW_r^i, a_k \right\rangle_{\mathcal{H}} \left\langle \int_0^t B_r(e_j) dW_r^j, a_k \right\rangle_{\mathcal{H}} \right).$$

Inspecting the last equality, by Proposition 2.10 we have that

$$\sum_{i=1}^{n} \left(\left\| \int_0^t B_r(e_i) dW_r^i \right\|_{\mathcal{H}}^2 - \int_0^t \| B_r(e_i) \|_{\mathcal{H}}^2 dr \right)$$

is a finite sum of martingales, so the process defined in (2.49) will itself be a martingale as well if we show that the same is true for

$$\sum_{k=1}^{\infty} \sum_{i \neq j} \left\langle \int_0^t B_r(e_i) dW_r^i, a_k \right\rangle_{\mathcal{H}} \left\langle \int_0^t B_r(e_j) dW_r^j, a_k \right\rangle_{\mathcal{H}}. \qquad (2.50)$$

We consider the above at first for each fixed k, rewriting it as

$$\sum_{i \neq j} \left(\int_0^t \langle B_r(e_j), a_k \rangle_{\mathcal{H}} dW_r^j \right) \left(\int_0^t \langle B_r(e_i), a_k \rangle_{\mathcal{H}} dW_r^i \right). \qquad (2.51)$$

Note that for each $i \neq j$,

$$\left(\int_0^t \langle B_r(e_j), a_k \rangle_{\mathcal{H}} dW_r^j \right) \left(\int_0^t \langle B_r(e_i), a_k \rangle_{\mathcal{H}} dW_r^i \right)$$

is the product of two independent real valued square integrable martingales, so is itself a martingale, and hence the finite sum given in (2.51) retains the martingale property. We emphasize again that the martingale property is always considered with respect to the fixed filtration $(\mathcal{F}_t)$ of our probability space.

We wish to show that the martingale property remains true in the limit of the infinite sum for (2.50). Convergence of the infinite sum is defined $\mathbb{P} - a.s.$, and it is sufficient to show that the convergence also holds in $L^1(\Omega; \mathbb{R})$ as this topology retains the martingale property. For this we show that the sequence is Cauchy in

$L^1(\Omega; \mathbb{R})$, taking the difference of the lth and mth terms to see that

$$
\mathbb{E}\left|\sum_{k=m+1}^{l}\sum_{i\neq j}\left(\int_0^t \langle B_r(e_j), a_k\rangle_{\mathcal{H}} dW_r^j\right)\left(\int_0^t \langle B_r(e_i), a_k\rangle_{\mathcal{H}} dW_r^i\right)\right|
$$

$$
\leq \sum_{k=m+1}^{l}\sum_{i\neq j}\mathbb{E}\left|\left(\int_0^t \langle B_r(e_j), a_k\rangle_{\mathcal{H}} dW_r^j\right)\left(\int_0^t \langle B_r(e_i), a_k\rangle_{\mathcal{H}} dW_r^i\right)\right|
$$

$$
\leq \frac{1}{2}\sum_{k=m+1}^{l}\sum_{i\neq j}\mathbb{E}\left[\left(\int_0^t \langle B_r(e_j), a_k\rangle_{\mathcal{H}} dW_r^j\right)^2 + \left(\int_0^t \langle B_r(e_i), a_k\rangle_{\mathcal{H}} dW_r^i\right)^2\right]
$$

$$
\leq n^2 \sum_{k=m+1}^{l}\sup_i \mathbb{E}\left(\int_0^t \langle B_r(e_i), a_k\rangle_{\mathcal{H}} dW_r^i\right)^2
$$

$$
\leq n^2 \sum_{k=m+1}^{l}\sup_i \mathbb{E}\int_0^t \langle B_r(e_i), a_k\rangle_{\mathcal{H}}^2 dr
$$

$$
\leq n^2 \sum_{k=m+1}^{\infty}\sup_i \mathbb{E}\int_0^t \langle B_r(e_i), a_k\rangle_{\mathcal{H}}^2 dr
$$

having used the Itô isometry. This is a monotone decreasing sequence in m, convergent to zero, hence the Cauchy property is shown so there exists a limit in $L^1(\Omega; \mathbb{R})$ which must agree with the $\mathbb{P} - a.s.$ limit (we can take a $\mathbb{P} - a.s.$ convergent subsequence from the $L^1(\Omega; \mathbb{R})$ convergence) and the martingale property of the process defined in (2.50), and hence (2.49) is shown. By applying Lemma 2.5, we deduce that $\left[\int_0^t B_r d\mathcal{W}_r\right]_t$ is the $L^1(\Omega; \mathbb{R})$ limit of the sequence

$$
\int_0^t \sum_{i=1}^{n} \|B_r(e_i)\|_{\mathcal{H}}^2 dr
$$

in n. Similarly to the analysis just conducted, we can show that this sequence is Cauchy in $L^1(\Omega; \mathbb{R})$ and its limit agrees with the $\mathbb{P} - a.s.$ limit, which is of course

$$
\int_0^t \|B_r\|_{\mathscr{L}^2(\mathfrak{U};\mathcal{H})}^2 dr
$$

taking the infinite sum through the integral with either Tonelli's Theorem (identifying the infinite sum as an integral with respect to the counting measure) or the Monotone Convergence Theorem. The proof is concluded.   □

We also have the analogous result to Proposition 2.11.

Proposition 2.20 *Let $B \in I_T^{\mathcal{H}}(\mathcal{W})$ and consider any sequence of partitions*

$$I_l := \left\{ 0 = t_0^l < t_1^l < \cdots < t_{k_l}^l = T \right\}$$

with $\max_j |t_j^l - t_{j-1}^l| \to 0$ as $l \to \infty$. Then for all $t \in [0, T]$, for any $\varepsilon > 0$,

$$\lim_{l \to \infty} \mathbb{P}\left(\left\{ \left| \sum_{t_{j+1}^l \leq t} \left\| \int_{t_j^l}^{t_{j+1}^l} B_r d\mathcal{W}_r \right\|_{\mathcal{H}}^2 - \int_0^t \|B_r\|_{\mathscr{L}^2(\mathfrak{U};\mathcal{H})}^2 dr \right| > \varepsilon \right\} \right) = 0.$$

(2.52)

Proof Following the method used in Proposition 2.11, we would again like to reduce this to a familiar case and extrapolate the result to the limit. We introduce a sequence of stopping times (τ^n) defined at every $n \in \mathbb{N}$ by

$$\tau^n := n \wedge \inf \left\{ t \in [0, T] : \left\| \int_0^t B_r d\mathcal{W}_r \right\|_{\mathcal{H}}^2 + \int_0^t \|B_r\|_{\mathscr{L}^2(\mathfrak{U};\mathcal{H})}^2 dr \geq n \right\}.$$

For every n we define the process

$$B_{\cdot}^n := B.\mathbb{1}_{\cdot \leq \tau^n}$$

and now look to show that

$$\lim_{l \to \infty} \mathbb{E}\left(\left| \sum_{t_{j+1}^l \leq t} \left\| \int_{t_j^l}^{t_{j+1}^l} B_r^n d\mathcal{W}_r \right\|_{\mathcal{H}}^2 - \int_0^t \|B_r^n\|_{\mathscr{L}^2(\mathfrak{U};\mathcal{H})}^2 dr \right| \right) = 0.$$

(2.53)

This is precisely in line with the method of Proposition 2.11. We have that

$$\mathbb{E}\left(\left| \sum_{t_{j+1}^l \leq t} \left\| \int_{t_j^l}^{t_{j+1}^l} B_r^n d\mathcal{W}_r \right\|_{\mathcal{H}}^2 - \int_0^t \|B_r^n\|_{\mathscr{L}^2(\mathfrak{U};\mathcal{H})}^2 dr \right| \right)$$

$$= \mathbb{E}\left(\left| \sum_{t_{j+1}^l \leq t} \left\| \int_{t_j^l}^{t_{j+1}^l} B_r^n d\mathcal{W}_r \right\|_{\mathcal{H}}^2 - \int_0^t \sum_{i=1}^{\infty} \|B_r^n(e_i)\|_{\mathcal{H}}^2 dr \right| \right)$$

$$= \mathbb{E}\left(\left| \sum_{t_{j+1}^l \leq t} \sum_{k=1}^{\infty} \left\langle \int_{t_j^l}^{t_{j+1}^l} B_r^n d\mathcal{W}_r, a_k \right\rangle_{\mathcal{H}}^2 - \int_0^t \sum_{i=1}^{\infty} \sum_{k=1}^{\infty} \langle B_r^n(e_i), a_k \rangle_{\mathcal{H}}^2 dr \right| \right)$$

$$= \mathbb{E}\left(\left|\sum_{t_{j+1}^l \le t} \sum_{k=1}^{\infty} \left(\int_{t_j^l}^{t_{j+1}^l} \langle B_r^n, a_k \rangle_{\mathcal{H}} d\mathcal{W}_r\right)^2 - \int_0^t \sum_{k=1}^{\infty} \sum_{i=1}^{\infty} \langle B_r^n(e_i), a_k \rangle_{\mathcal{H}}^2 dr\right|\right)$$

$$= \mathbb{E}\left(\left|\sum_{t_{j+1}^l \le t} \sum_{k=1}^{\infty} \left(\int_{t_j^l}^{t_{j+1}^l} \langle B_r^n, a_k \rangle_{\mathcal{H}} d\mathcal{W}_r\right)^2 - \int_0^t \sum_{k=1}^{\infty} \left\|\langle B_r^n(e_i), a_k \rangle_{\mathcal{H}}\right\|_{\mathscr{L}^2(\mathfrak{U};\mathbb{R})}^2 dr\right|\right)$$

$$\le \sum_{k=1}^{\infty} \mathbb{E}\left(\left|\sum_{t_{j+1}^l \le t} \left(\int_{t_j^l}^{t_{j+1}^l} \langle B_r^n, a_k \rangle_{\mathcal{H}} d\mathcal{W}_r\right)^2 - \int_0^t \left\|\langle B_r^n(e_i), a_k \rangle_{\mathcal{H}}\right\|_{\mathscr{L}^2(\mathfrak{U};\mathbb{R})}^2 dr\right|\right)$$

having applied Proposition 2.16 and the Dominated Convergence Theorem to take the infinite sum in k through the time integral and expectation. From Propositions 2.16 and 2.18, then $\int_0^\cdot \langle B_r^n, a_k \rangle_{\mathcal{H}} d\mathcal{W}_r$ belongs to $\mathcal{M}_c^2$, with quadratic variation $\int_0^t \left\|\langle B_r(e_i), a_k \rangle_{\mathcal{H}}\right\|_{\mathscr{L}^2(\mathfrak{U};\mathbb{R})}^2 dr$ coming from Proposition 2.19. Just as we used in Proposition 2.11, we have that for each fixed $k \in \mathbb{N}$,

$$\lim_{l \to \infty} \mathbb{E}\left(\left|\sum_{t_{j+1}^l \le t} \left(\int_{t_j^l}^{t_{j+1}^l} \langle B_r^n, a_k \rangle_{\mathcal{H}} d\mathcal{W}_r\right)^2 - \int_0^t \left\|\langle B_r^n(e_i), a_k \rangle_{\mathcal{H}}\right\|_{\mathscr{L}^2(\mathfrak{U};\mathbb{R})}^2 dr\right|\right) = 0$$

so it is sufficient to justify the interchange of limit in l and summation in k. This follows identically to the justification in Proposition 2.11, appealing this time to the Itô isometry Proposition 2.15.   □

Chapter 3
Stochastic Differential Equations in Infinite Dimensions

In this chapter we establish a framework for the study of stochastic partial differential equations (SPDEs), which are evolution equations involving integration of the form introduced in the previous chapter. Through this framework we define notions of solutions for an abstract SPDE, incorporating both *unbounded* (in the sense of differential operators) and *Stratonovich* noise. One main result is the rigorous conversion between Itô and Stratonovich forms under an unbounded noise.

3.1 The Stratonovich Integral

We look at first to define the Stratonovich Integral with respect to a one dimensional martingale, before doing so with respect to a Cylindrical Brownian Motion. Recall that the Itô stochastic integral was constructed in Definitions 2.10, 2.20, and 2.22.

Definition 3.1 For $M \in \bar{\mathcal{M}}_c^2$ and $\Psi \in \mathcal{I}_M^{\mathcal{H}} \cap \bar{\mathcal{M}}_c^2(\mathcal{H})$, the Stratonovich stochastic integral is defined as

$$\int_0^t \Psi_s \circ dM_s := \int_0^t \Psi_s dM_s + \frac{1}{2}[\Psi, M]_t.$$

Definition 3.2 For $B \in \mathcal{I}^{\mathcal{H}}(\mathcal{W})$ such that $B_{e_i} \in \bar{\mathcal{M}}_c^2(\mathcal{H})$ for every e_i and the limit

$$\sum_{i=1}^{\infty}[B_{e_i}, W^i]_t \tag{3.1}$$

© The Author(s), under exclusive license to Springer Nature Switzerland AG 2024
D. Goodair, D. Crisan, *Stochastic Calculus in Infinite Dimensions and SPDEs*,
SpringerBriefs in Mathematics, https://doi.org/10.1007/978-3-031-69586-5_3

is well defined in $L^2(\Omega; \mathcal{H})$, the Stratonovich stochastic integral is defined as

$$\int_0^t B(s) \circ d\mathcal{W}_s := \sum_{i=1}^{\infty} \left(\int_0^t B_{e_i}(s) dW_s^i + \frac{1}{2}[B_{e_i}, W^i]_t \right),$$

where the limit is taken in $L^2(\Omega; \mathcal{H})$. The class of such processes will be denoted $\mathcal{I}_\circ^{\mathcal{H}}(\mathcal{W})$.

It will be necessary to extend this definition to processes $B \in \bar{\mathcal{I}}^{\mathcal{H}}(\mathcal{W})$, but we encounter more technical issues in trying to construct a sequence of stopping times such that the stopped process belongs to $\mathcal{I}_\circ^{\mathcal{H}}(\mathcal{W})$. We find it simplest to give the definition below.

Definition 3.3 Suppose that there exists a sequence of stopping times (τ_n) which are $\mathbb{P} - a.s.$ monotone increasing and convergent to infinity such that:

1. For every n, the process

$$B^n(\cdot) := B(\cdot) \mathbb{1}_{\cdot \leq \tau^n}$$

belongs to $\mathcal{I}^{\mathcal{H}}(\mathcal{W})$.
2. For every n and i, the process

$$B_{e_i}^{\tau_n}(\cdot) := B_{e_i}(\cdot \wedge \tau^n)$$

belongs to $\bar{\mathcal{M}}_c^2(\mathcal{H})$.
3. The limit

$$\sum_{i=1}^{\infty}[B_{e_i}^{\tau_n}, W^i]_t$$

is well defined in $L^2(\Omega; \mathcal{H})$.

Then the Stratonovich stochastic integral is defined at any $t \geq 0$ as

$$\int_0^t B(s) \circ d\mathcal{W}_s := \lim_{n \to \infty} \left(\sum_{i=1}^{\infty} \left(\int_0^t B_{e_i}^n(s) dW_s^i + \frac{1}{2}[B_{e_i}^{\tau_n}, W^i]_t \right) \right),$$

where the limit is taken $\mathbb{P} - a.s.$ in $\mathcal{H}$, and the infinite sum in $L^2(\Omega; \mathcal{H})$. The class of such processes will be denoted $\bar{\mathcal{I}}_\circ^{\mathcal{H}}(\mathcal{W})$.

This definition is of course completely analogous to the localization procedure used in the previous constructions, for example, Definition 2.22, except we postulate in the first instance the existence of the localizing sequence (τ_n). There does not appear to be a simple canonical way to construct such a sequence of localizing times whereby each $B_{e_i}^{\tau_n} \in \bar{M}_c^2(\mathcal{H})$ if one assumes only local martingality of these component processes, and hence we assume the required properties explicitly.

3.2   Strong Solutions in the Abstract Framework

We now establish a framework in which to formulate SPDEs, posed for a quartet of embedded Hilbert Spaces

$$V \hookrightarrow H \hookrightarrow U \hookrightarrow X$$

where the embedding is meant as a continuous linear injection. Each Hilbert Space plays a role in defining solutions, and this structure of embeddings allows us to consider operators which are not bounded on a given Hilbert Space (particularly nonlinear and differential operators in applications). We introduce at first the Itô SPDE

$$\Psi_t = \Psi_0 + \int_0^t Q\Psi_s ds + \int_0^t \mathcal{G}\Psi_s dW_s, \qquad (3.2)$$

where $\mathcal{W}$ continues to be a Cylindrical Brownian Motion over $\mathfrak{U}$ relative to our fixed filtered probability space $(\Omega, \mathcal{F}, (\mathcal{F}_t), \mathbb{P})$ with representation (2.36). We impose now some conditions on the operators relative to these spaces. To do so we define the general map $\tilde{K} = \tilde{K}_{c,p,q} : H \to \mathbb{R}$ by

$$\tilde{K}(\phi) := c \left(1 + \|\phi\|_U^p + \|\phi\|_H^q\right)$$

for any constants c, p, q independent of ϕ.

Assumption 3.1 $Q : V \to U$ is measurable, and there exists a map $\tilde{K} = \tilde{K}_{c,p,q}$ such that for any $\phi \in V$,

$$\|Q\phi\|_U \leq \tilde{K}(\phi)[1 + \|\phi\|_V^2].$$

Assumption 3.2 $\mathcal{G}$ is understood as a measurable operator

$$\mathcal{G} : U \to \mathcal{L}^2(\mathfrak{U}; X), \qquad \mathcal{G}|_H : H \to \mathcal{L}^2(\mathfrak{U}; U), \qquad \mathcal{G}|_V : V \to \mathcal{L}^2(\mathfrak{U}; H)$$

defined over $\mathfrak{U}$ by its action on the basis vectors

$$\mathcal{G}(\cdot, e_i) := \mathcal{G}_i(\cdot).$$

Each $\mathcal{G}_i$ is linear, and there exist constants c_i such that for all $\phi \in V$, $\psi \in H$, $\eta \in U$:

$$\|\mathcal{G}_i \eta\|_X \le c_i \|\eta\|_U , \qquad \|\mathcal{G}_i \psi\|_U \le c_i \|\psi\|_H , \qquad \|\mathcal{G}_i \phi\|_H \le c_i \|\phi\|_V$$

$$\sum_{i=1}^{\infty} c_i^2 < \infty.$$

Here $\mathcal{G}_i$ is assumed linear for the purposes of the Stratonovich conversion in Sect. 3.4 but would not otherwise be necessary. It is worth clarifying how $\mathcal{G}$ is defined over $\mathfrak{U}$: Fix a $\phi \in V$ and consider $\mathcal{G}(\phi, \cdot) : \mathfrak{U} \to H$ (the arguments here apply for the larger spaces as well). Any $\alpha \in \mathfrak{U}$ has the representation

$$\sum_{i=1}^{\infty} \langle \alpha, e_i \rangle_{\mathfrak{U}} e_i,$$

where

$$\sum_{i=1}^{\infty} \langle \alpha, e_i \rangle_{\mathfrak{U}}^2 < \infty.$$

Then

$$\mathcal{G}(\phi, \cdot) : \alpha \mapsto \sum_{i=1}^{\infty} \langle \alpha, e_i \rangle_{\mathfrak{U}} \mathcal{G}_i \phi$$

is well defined as an element of H. This is justified by showing that the sequence of partial sums is Cauchy in H: Note that from Cauchy–Schwarz,

$$\left\| \sum_{i=m}^{n} \langle \alpha, e_i \rangle_{\mathfrak{U}} \mathcal{G}_i \phi \right\|_H^2 \le \left(\sum_{i=m}^{n} \langle \alpha, e_i \rangle_{\mathfrak{U}}^2 \right) \left(\sum_{i=m}^{n} \|\mathcal{G}_i \phi\|_H^2 \right)$$

$$\le \left(\sum_{i=m}^{\infty} \langle \alpha, e_i \rangle_{\mathfrak{U}}^2 \right) \left(\sum_{i=m}^{\infty} c_i^2 \|\phi\|_V^2 \right),$$

which approaches zero as $m \to \infty$ as the sums are finite. We introduce now the first notion of our strong solutions.

Definition 3.4 Let $\Psi_0 : \Omega \to H$ be $\mathcal{F}_0$-measurable. A pair (Ψ, τ) where τ is a $\mathbb{P} - a.s.$ positive stopping time and Ψ, is a process such that for $\mathbb{P} - a.e.$ ω, $\Psi.(\omega) \in C([0, T]; H)^1$ and $\Psi.(\omega)\mathbb{1}._{\le \tau(\omega)} \in L^2([0, T]; V)$ with $\Psi.\mathbb{1}._{\le \tau}$

---

[1] We make the definition for a continuous process as this property is desirable, but it is not necessary to make the definition and we may not always have continuity. We shall see that one only requires $L^{\infty}([0, T]; H)$ regularity to construct the integrals in (3.2), and hence we face no issue if and when the continuity is relieved.

progressively measurable in V is said to be a local strong solution of the equation (3.2) if the identity

$$\Psi_t = \Psi_0 + \int_0^{t \wedge \tau} Q\Psi_s ds + \int_0^{t \wedge \tau} \mathcal{G}\Psi_s dW_s \qquad (3.3)$$

holds $\mathbb{P} - a.s.$ in U for all $t \in [0, T]$.

Remark 3.1 Note that if (Ψ, τ) is a local strong solution of equation (3.2), then $\Psi. = \Psi._{\wedge \tau}$ due to the integral representation (3.3). That is to say, the solution is automatically stopped at the local time of existence. Moreover the progressive measurability condition on $\Psi.\mathbb{1}._{\leq \tau}$ may look a little suspect as Ψ_0 itself may only belong to H and not V making it impossible for $\Psi.\mathbb{1}._{\leq \tau}$ to be even adapted in V. We have mildly abusing notation here; what we really ask is that there exists a process Φ which is progressively measurable in V and such that $\Phi. = \Psi.\mathbb{1}._{\leq \tau}$ almost surely over the product space $\Omega \times [0, T]$ with product measure $\mathbb{P} \times \lambda$. In particular, the processes Φ and Ψ may disagree at time zero.

Let us take a few moments to process this definition and ensure that the integrals make sense with the given regularity of the solution and the operators Q, $\mathcal{G}$. The time integral in (3.3) is well defined in U as a Bochner Integral: First of all $\Psi.(\omega)\mathbb{1}._{\leq \tau(\omega)} \in L^2([0, T]; V)$ for $\mathbb{P} - a.e.$ ω so $\Psi.(\omega)\mathbb{1}._{\leq \tau(\omega)} : [0, t] \to V$ is measurable and hence $Q(\Psi.(\omega)\mathbb{1}._{\leq \tau(\omega)}) : [0, t] \to U$ is measurable from Assumption 3.1. Moreover, the mapping $Q(\Psi.(\omega)\mathbb{1}._{\leq \tau(\omega)})\mathbb{1}._{\leq \tau(\omega)} : [0, t] \to U$ is again measurable, and we have that

$$\int_0^{t \wedge \tau(\omega)} Q(\Psi_s(\omega)) \, ds = \int_0^t Q(\Psi_s(\omega)\mathbb{1}_{s \leq \tau(\omega)}) \, \mathbb{1}_{s \leq \tau(\omega)} ds$$

so the required measurability in order to define the integral is satisfied. In this vein, we have that

$$\int_0^{t \wedge \tau(\omega)} \|Q(\Psi_s(\omega))\|_U \, ds \leq \int_0^{t \wedge \tau(\omega)} \tilde{K}(\Psi_s(\omega)) \left[1 + \|\Psi_s(\omega)\|_V^2\right] ds$$

$$\leq \sup_{s \in [0, t \wedge \tau(\omega)]} \left[\tilde{K}(\Psi_s(\omega))\right] \int_0^{t \wedge \tau(\omega)} 1 + \|\Psi_s(\omega)\|_V^2 \, ds$$

$$< \infty$$

employing Assumption 3.1 again and using the regularity specified by the local solution to deduce finiteness, which justifies that the integral is well defined.

As for the stochastic integral in (3.3), this is well defined in the sense of Definition 2.22 in H (which then embeds into U), though it is done so formally via the process Φ stipulated in Remark 3.1. We understand again that

$$\int_0^{t\wedge\tau} \mathcal{G}\Psi_s dW_s = \int_0^t \mathcal{G}(\Psi_s \mathbb{1}_{s\leq\tau})\mathbb{1}_{s\leq\tau} dW_s$$

should the integral be well defined, and so we actually define the integral by

$$\int_0^{t\wedge\tau} \mathcal{G}\Psi_s dW_s := \int_0^t \mathcal{G}\Phi_s \mathbb{1}_{s\leq\tau} dW_s.$$

The progressive measurability of $\mathcal{G}\Phi\mathbb{1}_{\cdot\leq\tau} : [0, t] \times \Omega \to \mathscr{L}^2(\mathfrak{U}; H)$ is immediate from the measurability assumption of Assumption 3.2, the same arguments above for the time integral and the progressive measurability of $\mathbb{1}_{\cdot\leq\tau}$ coming from the definition of the stopping time. Similarly we have that

$$\int_0^{t\wedge\tau} \sum_{i=1}^{\infty} \|\mathcal{G}_i(\Phi_s(\omega))\|_H^2 ds = \int_0^{t\wedge\tau} \sum_{i=1}^{\infty} \|\mathcal{G}_i(\Psi_s(\omega)\mathbb{1}_{s\leq\tau})\|_H^2 ds < \infty$$

$\mathbb{P} - a.s.$, using again Assumption 3.2, validating that the stochastic integral is well defined as a local martingale in H and thus in U from the continuous embedding. It should be observed that moment estimates are not assumed for our solution (or even the initial condition), so we cannot say that the stochastic integral is genuinely a square integrable martingale. We clarify another potential source of ambiguity with the following lemma, around whether there could be a meaningful dependence of the set of full probability on t used to satisfy (3.3) of Definition 3.4. The lemma justifies that it is sufficient to check the identity (3.3) at each $t \in [0, T]$ $\mathbb{P} - a.s..$.

Lemma 3.1 *Suppose that there exists a set $\bar{\Omega} \subset \Omega$, $\mathbb{P}(\bar{\Omega}) = 1$, and a pair (Ψ, τ) where τ is positive on $\bar{\Omega}$ and Ψ is a process such that for every $\omega \in \bar{\Omega}$, $\Psi_{\cdot}(\omega) \in L^{\infty}([0, T]; H)$ and $\Psi_{\cdot}(\omega)\mathbb{1}_{\cdot\leq\tau(\omega)} \in L^2([0, T]; V)$. In addition, suppose that for every $t \in [0, T]$ there exists a set $\Omega_t \subset \Omega$, $\mathbb{P}(\Omega_t) = 1$ whereby the identity (3.3) holds on Ω_t. Then there exists a set $\check{\Omega} \subset \Omega$, $\mathbb{P}(\check{\Omega}) = 1$, and a process $\check{\Psi}$ such that for every $\omega \in \check{\Omega}$, $\check{\Psi}_{\cdot}(\omega) \in L^{\infty}([0, T]; H)$ and $\check{\Psi}_{\cdot}(\omega)\mathbb{1}_{\cdot\leq\tau(\omega)} \in L^2([0, T]; V)$, where $\check{\Psi}$ satisfies the identity*

$$\check{\Psi}_t = \Psi_0 + \int_0^{t\wedge\tau} Q\check{\Psi}_s ds + \int_0^{t\wedge\tau} \mathcal{G}\check{\Psi}_s dW_s \qquad (3.4)$$

for all $\omega \in \check{\Omega}$ and $t \in [0, T]$. Furthermore, if we assume that for every $\omega \in \bar{\Omega}$, $\Psi_{\cdot}(\omega) \in C([0, T]; H)$, then we can choose $\check{\Omega}$ and $\check{\Psi}$ such that $\Psi_t(\omega) = \check{\Psi}_t(\omega)$ for all $\omega \in \check{\Omega}$, $t \in [0, T]$.

Proof We consider all rational times (t_k) in $[0, T]$ and construct

$$\check{\Omega} = \bar{\Omega} \cap \bigcap_k \Omega_{t_k}$$

as well as the process $\check{\Psi}$ on $\check{\Omega} \times [0, T]$ by

$$\check{\Psi}_t = \Psi_0 + \int_0^{t \wedge \tau} Q\Psi_s ds + \int_0^{t \wedge \tau} \mathcal{G}\Psi_s dW_s.$$

Clearly, $\mathbb{P}(\check{\Omega}) = 1$ and $\Psi_{t_k}(\omega) = \check{\Psi}_{t_k}(\omega)$ for all $\omega \in \check{\Omega}$, so $\Psi_\cdot(\omega)$ and $\check{\Psi}_\cdot(\omega)$ are equal $\lambda - a.s.$ on $[0, T]$, and hence $\check{\Psi}_\cdot(\omega) \in L^\infty([0, T]; H)$ and $\check{\Psi}_\cdot(\omega)\mathbb{1}_{\cdot \le \tau(\omega)} \in L^2([0, T]; V)$. Due to this equivalence, then on $\check{\Omega}$,

$$\int_0^{t \wedge \tau} Q\Psi_s ds = \int_0^{t \wedge \tau} Q\check{\Psi}_s ds, \qquad \int_0^{t \wedge \tau} \mathcal{G}\Psi_s dW_s = \int_0^{t \wedge \tau} \mathcal{G}\check{\Psi}_s dW_s$$

so the identity (3.4) follows. With the additional assumption of continuity then we not only have that for all $\omega \in \check{\Omega}$, $\Psi_\cdot(\omega)$ and $\check{\Psi}_\cdot(\omega)$ are equal $\lambda - a.s.$ on $[0, T]$, but also that $\Psi_\cdot(\omega) \in C([0, T]; H)$ and hence $C([0, T]; U)$. From the identity (3.4), then $\check{\Psi}_\cdot(\omega) \in C([0, T]; U)$ as well, so $\Psi_\cdot(\omega)$ and $\check{\Psi}_\cdot(\omega)$ are equal $\lambda - a.s.$ on $[0, T]$ and are both continuous (into U) so must in fact be equal at all times $t \in [0, T]$. $\square$

We treat the local solution here to deal with the additional technicalities which arise from accounting for the stopping time. Global solutions are similarly defined however.

Definition 3.5 Let $\Psi_0 : \Omega \to H$ be $\mathcal{F}_0$-measurable. A process Ψ such that for $\mathbb{P} - a.e.$ ω, $\Psi_\cdot(\omega) \in C([0, T]; H) \cap L^2([0, T]; V)$ with Ψ progressively measurable in V is said to be a **strong solution** of the equation (3.2) if the identity (3.2) holds $\mathbb{P} - a.s.$ in U for all $t \in [0, T]$.

The solution is global in the sense that the identity holds on the given interval $[0, T]$, where T was arbitrary. One could certainly argue that a true global solution should be defined on $[0, \infty)$, inviting the question as to when these notions are equivalent; one answer is in the case where solutions are unique, ensuring that the solution on $[0, T + 1]$ is an extension of that on $[0, T]$ and as such can be extended to $[0, \infty)$. Uniqueness is considered in the next section.

3.3 Uniqueness and Maximality

In this section we introduce the notion of uniqueness used in our solutions, as well as passage from local strong solutions to a stopping time which is *maximal*. We first state the key definitions.

Definition 3.6 A local strong solution (Ψ, τ) of the equation (3.2) is said to be unique if for any other such local strong solution (Φ, θ), then

$$\mathbb{P}(\{\omega \in \Omega : \Psi_t(\omega) = \Phi_t(\omega) \quad \forall t \in [0, T \wedge \tau(\omega) \wedge \theta(\omega)]\}) = 1.$$

Remark 3.2 The uniqueness is in the sense of indistinguishability, Definition 2.15, up until the minimum of the stopping times and $T > 0$.

Definition 3.7 A pair (Ψ, Θ) such that there exists a sequence of stopping times (θ_j) which are $\mathbb{P} - a.s.$ monotone increasing and convergent to Θ, whereby $(\Psi_{\cdot \wedge \theta_j}, \theta_j)$ is a local strong solution of the equation (3.2) for each j, is said to be a maximal strong solution of the equation (3.2) if for any other pair (Φ, Γ) with this property, then $\Theta \le \Gamma \, \mathbb{P} - a.s.$ implies $\Theta = \Gamma \, \mathbb{P} - a.s..$

Definition 3.8 A maximal strong solution (Ψ, Θ) of the equation (3.2) is said to be unique if for any other such solution (Φ, Γ), then $\Theta = \Gamma \, \mathbb{P} - a.s.$ and

$$\mathbb{P}\left(\{\omega \in \Omega : \Psi_t(\omega) = \Phi_t(\omega) \quad \forall t \in [0, T \wedge \Theta(\omega))\}\right) = 1.$$

The remainder of this section is devoted to the following result.

Theorem 3.3 *Define a "local regular solution" of the equation (3.2) to be a local strong solution* (Ψ, τ)*, with the alteration that the progressive measurability is either satisfied for the genuine process* $\Psi . \mathbb{1}_{\cdot \le \tau}$ *in V (and thus, not a version of it) or of* Ψ *in H. Suppose that there exists a local regular solution of the equation (3.2) and that any local regular solution is unique. Then there exists a unique maximal regular solution of the equation (3.2).*

To this end, we state and prove an important result in both this context and future ones.

Proposition 3.1 *Let* (A_k) *be a partition of* Ω*, which is to say that* $A_k \cap A_j = \emptyset$ *when* $k \ne j$ *and* $\bigcup_k A_k = \Omega$*, where each* $A_k \in \mathcal{F}_0$*. Consider a sequence* (Ψ_0^k)*, each* $\Psi_0^k : \Omega \to H \, \mathcal{F}_0$*-measurable, and let* $((\Psi^k, \tau^k))$ *be local strong solutions of the equation (3.2) corresponding to* (Ψ_0^k)*. Then the pair* (Ψ, τ) *defined by*

$$\Psi = \sum_{k=1}^{\infty} \Psi^k \mathbb{1}_{A_k}, \qquad \tau = \sum_{k=1}^{\infty} \tau^k \mathbb{1}_{A_k}$$

is a local strong solution of the equation (3.2) for the initial condition

$$\Psi_0 = \sum_{k=1}^{\infty} \Psi_0^k \mathbb{1}_{A_k}.$$

Proof We first note that the infinite sums in defining (Ψ, τ) and Ψ_0 are given by a single term for any ω, so the limits are trivially defined $\mathbb{P} - a.s.$ in $C([0, T]; U)$, $\mathbb{R}$, and $\mathbb{R}$, respectively. As each $A_k \in \mathcal{F}_0$, then $\mathbb{1}_{A_k}$ is $\mathcal{F}_0$-measurable and hence each $\tau^k \mathbb{1}_{A_k}$ remains a stopping time so their $\mathbb{P} - a.s.$ limit τ is again a stopping time. It is clear that for $\mathbb{P} - a.e.$ ω, $(\Psi(\omega), \tau(\omega)) = (\Psi^k(\omega), \tau^k(\omega))$ for some k, and hence pathwise properties of the local strong solution are retained. That is, for $\mathbb{P} - a.e.$ ω, $\Psi_{\cdot}(\omega) \in L^{\infty}([0, T]; H)$ and $\Psi_{\cdot}(\omega)\mathbb{1}_{\cdot \le \tau(\omega)} \in L^2([0, T]; V)$. We next verify the identity (3.3), where progressive measurability (thus ensuring that the stochastic integral is indeed well defined) is shown afterward. As the (A_k) partition Ω, it is sufficient to show that for every $t \in [0, T]$, the identity

$$\mathbb{1}_{A_k} \Psi_t = \mathbb{1}_{A_k} \Psi_0 + \mathbb{1}_{A_k} \int_0^{t \wedge \tau} Q\Psi_s ds + \mathbb{1}_{A_k} \int_0^{t \wedge \tau} G\Psi_s dW_s \qquad (3.5)$$

holds $\mathbb{P} - a.s.$ in U. We now look to reduce each term, seeing that

$$\mathbb{1}_{A_k} \Psi_t = \mathbb{1}_{A_k} \Psi_t^k$$

$$\mathbb{1}_{A_k} \Psi_0 = \mathbb{1}_{A_k} \Psi_0^k$$

$$\mathbb{1}_{A_k} \int_0^{t \wedge \tau} Q\Psi_s ds = \mathbb{1}_{A_k} \int_0^{t \wedge \tau^k} Q\Psi_s^k ds$$

holds $\mathbb{P} - a.s.$ in U, where the last equality is justified by

$$\mathbb{1}_{A_k} \int_0^{t \wedge \tau} Q\Psi_s ds = \left(\mathbb{1}_{A_k} \right)^2 \int_0^{t \wedge \tau} Q\Psi_s ds = \mathbb{1}_{A_k} \int_0^t \mathbb{1}_{A_k} \mathbb{1}_{s \leq \tau} Q\Psi_s ds$$

$$= \mathbb{1}_{A_k} \int_0^t \mathbb{1}_{A_k} \mathbb{1}_{A_k \cap \{s \leq \tau\}} Q(\Psi_s \mathbb{1}_{A_k}) ds$$

$$= \mathbb{1}_{A_k} \int_0^t \mathbb{1}_{A_k} \mathbb{1}_{A_k \cap \{s \leq \tau^k\}} Q(\Psi_s^k \mathbb{1}_{A_k}) ds$$

$$= \mathbb{1}_{A_k} \int_0^{t \wedge \tau^k} Q(\Psi_s^k) ds$$

simply reversing the procedure to obtain the last line. We can apply the same arguments for the stochastic integral to obtain that

$$\mathbb{1}_{A_k} \int_0^{t \wedge \tau} G\Psi_s dW_s = \mathbb{1}_{A_k} \int_0^{t \wedge \tau^k} G\Psi_s^k dW_s \qquad (3.6)$$

though we now have to be careful in justifying passage of $\mathbb{1}_{A_k}$ through the stochastic integral; as $\mathbb{1}_{A_k}$ is $\mathcal{F}_0$-measurable, then we can apply Proposition 2.17. Therefore, to verify (3.5), it is sufficient to demonstrate that

$$\mathbb{1}_{A_k} \Psi_t^k = \mathbb{1}_{A_k} \Psi_0^k + \mathbb{1}_{A_k} \int_0^{t \wedge \tau^k} Q\Psi_s^k ds + \mathbb{1}_{A_k} \int_0^{t \wedge \tau^k} G\Psi_s^k dW_s,$$

but this follows from the fact that (Ψ^k, τ^k) is a local strong solution for the initial condition Ψ_0^k. To conclude that (Ψ, τ) is a local strong solution for the initial condition Ψ_0, it only remains to show that $\Psi . \mathbb{1}_{\leq \tau}$ is progressively measurable in V, which we deduce from

$$\Psi . \mathbb{1}_{.\leq\tau} = \left(\sum_{k=1}^{\infty} \Psi^k . \mathbb{1}_{A_k}\right) \mathbb{1}_{.\leq\tau} = \sum_{k=1}^{\infty} \Psi^k . \mathbb{1}_{.\leq\tau^k} \mathbb{1}_{A_k},$$

where the limit can be taken pointwise almost everywhere over the product space $\Omega \times [0, T]$ in V^2 (again for each fixed element of this product space, the infinite sum is in reality just a single term). From the progressive measurability of solutions, we have that for each N, the process $\sum_{k=1}^{N} \Psi^k . \mathbb{1}_{.\leq\tau^k} \mathbb{1}_{A_k}$ is progressively measurable (this is undisturbed by the $\mathcal{F}_0$-measurable $\mathbb{1}_{A_k}$) and hence for each fixed S is measurable as a mapping $\Omega \times [0, S] \to V$ where we equip $\Omega \times [0, S]$ with the sigma algebra $\mathcal{F}_S \times \mathcal{B}([0, S])$. The pointwise $\mathbb{P} \times \lambda - a.e.$ limit as $N \to \infty$ preserves the measurability which concludes the argument that (Ψ, τ) is a local strong solution of the equation (3.2) for the initial condition Ψ_0. $\square$

With this in place we set up notation for proving Theorem 3.3. We define $\mathcal{X}$ as the set of all stopping times σ such that there exists a process Ψ for which (Ψ, σ) is a local regular solution. We also define $\mathcal{Y}$ as the set of all stopping times given by the $\mathbb{P} - a.s.$ limit of monotone increasing elements of $\mathcal{X}$. We prove the existence of a maximal solution by showing that the maximum of any two elements of $\mathcal{X}$ is again in $\mathcal{X}$, a property which we use for the sequences in $\mathcal{X}$ to bound sequences in $\mathcal{Y}$ which then enables an application of Zorn's Lemma to deduce a maximal element. Of course, an assumption of Theorem 3.3 is that $\mathcal{X}$ is non-empty.

Lemma 3.2 *Under the assumptions of Theorem 3.3, for any* $\sigma_1, \sigma_2 \in \mathcal{X}$, *we have that* $\sigma_1 \vee \sigma_2 \in \mathcal{X}$.

Proof Our proof adapts the methodology of Proposition 3.1, but we establish the idea first. By definition of $\mathcal{X}$ we have that (Ψ^1, σ_1) and (Ψ^2, σ_2) are local regular solutions for some processes Ψ^1, Ψ^2, for the initial condition $\Psi_0^1 = \Psi_0^2 = \Psi_0$. The uniqueness of solutions establishes that $\Psi^1_{.\wedge\sigma_1\wedge\sigma_2}$ and $\Psi^2_{.\wedge\sigma_1\wedge\sigma_2}$ are indistinguishable. Our approach is to construct a new solution Ψ, which agrees with these indistinguishable processes up until $\sigma_1 \wedge \sigma_2$ and then extends to σ_1 as Ψ^1 if $\sigma_1 > \sigma_2$ and similarly for the reverse. One can construct the process Ψ and stopping time $\sigma = \sigma_1 \vee \sigma_2$ by

$$\Psi := \Psi^1 \mathbb{1}_{\sigma_1 \geq \sigma_2} + \Psi^2 \mathbb{1}_{\sigma_1 < \sigma_2}$$

$$\sigma := \sigma_1 \mathbb{1}_{\sigma_1 \geq \sigma_2} + \sigma_2 \mathbb{1}_{\sigma_1 < \sigma_2}.$$

A comparison with Proposition 3.1 is immediately noted through the sets $A_1 = \{\omega \in \Omega : \sigma_1(\omega) \geq \sigma_2(\omega)\}$ and $A_2 = \{\omega \in \Omega : \sigma_1(\omega) < \sigma_2(\omega)\}$, though these sets

---

[2] We note that strictly, this limit is taken for the representation Φ described in Remark 3.1. Only local strong solutions, not local regular solutions as considered in Theorem 3.3, are required in this result.

are not $\mathcal{F}_0$-measurable and hence we cannot apply the result directly. To show that (Ψ, σ) is indeed a local regular solution, we rewrite Ψ through

$$
\begin{aligned}
\Psi_\cdot &= \left(\Psi^1_\cdot - \Psi^1_{\cdot \wedge \sigma_1 \wedge \sigma_2} + \Psi^1_{\cdot \wedge \sigma_1 \wedge \sigma_2} \right) \mathbb{1}_{\sigma_1 \geq \sigma_2} \\
&\quad + \left(\Psi^2_\cdot - \Psi^1_{\cdot \wedge \sigma_1 \wedge \sigma_2} + \Psi^1_{\cdot \wedge \sigma_1 \wedge \sigma_2} \right) \mathbb{1}_{\sigma_1 < \sigma_2} \\
&= \Psi^1_{\cdot \wedge \sigma_1 \wedge \sigma_2} + \left(\Psi^1_\cdot - \Psi^1_{\cdot \wedge \sigma_1 \wedge \sigma_2} \right) \mathbb{1}_{\sigma_1 \geq \sigma_2} + \left(\Psi^2_\cdot - \Psi^1_{\cdot \wedge \sigma_1 \wedge \sigma_2} \right) \mathbb{1}_{\sigma_1 < \sigma_2} \\
&= \Psi^1_{\cdot \wedge \sigma_1 \wedge \sigma_2} + \left(\Psi^1_\cdot - \Psi^1_{\cdot \wedge \sigma_1 \wedge \sigma_2} \right) \mathbb{1}_{\sigma_1 \geq \sigma_2} + \left(\Psi^2_\cdot - \Psi^2_{\cdot \wedge \sigma_1 \wedge \sigma_2} \right) \mathbb{1}_{\sigma_1 < \sigma_2}. \quad (3.7)
\end{aligned}
$$

The purpose of this becomes clear when we look at the role of the $\mathcal{F}_0$-measurability in the proof of Proposition 3.1. First of all is for progressive measurability. We recall the definition of the stopped sigma algebra

$$
\mathcal{F}_{\sigma_k} = \{ A \in \mathcal{F} : \forall t \in [0, T], \, A \cap \{\sigma_k \leq t\} \in \mathcal{F}_t \}
$$

and the property that $\mathbb{1}_{\sigma_1 \geq \sigma_2}$ is $\mathcal{F}_{\sigma_1 \wedge \sigma_2}$-measurable. The same is true of $\mathbb{1}_{\sigma_1 < \sigma_2}$. Let us consider the case of progressive measurability in V, where the H case follows more simply. It is required to show that $\Psi_\cdot \mathbb{1}_{\cdot \leq \sigma_1 \vee \sigma_2}$ is progressively measurable in V. We show that this is true in each of the three terms in (3.7). In the first, we recall that a progressively measurable process stopped at a stopping time is again progressively measurable;[3] thus, $\Psi^1_{\cdot \wedge \sigma_1 \wedge \sigma_2} \mathbb{1}_{\cdot \wedge \sigma_1 \wedge \sigma_2 \leq \sigma_1} = \Psi^1_{\cdot \wedge \sigma_1 \wedge \sigma_2}$ is progressively measurable. So too, then, is its product with the real valued progressively measurable $\mathbb{1}_{\cdot \leq \sigma_1 \vee \sigma_2}$. For the following term in (3.7), note that

$$
\left(\Psi^1_\cdot - \Psi^1_{\cdot \wedge \sigma_1 \wedge \sigma_2} \right) \mathbb{1}_{\sigma_1 \geq \sigma_2} \mathbb{1}_{\cdot \leq \sigma_1 \vee \sigma_2} = \left(\Psi^1_\cdot - \Psi^1_{\cdot \wedge \sigma_2} \right) \mathbb{1}_{\sigma_1 \geq \sigma_2} \mathbb{1}_{\cdot \leq \sigma_1}
$$

$$
= \left(\Psi^1_\cdot \mathbb{1}_{\cdot \leq \sigma_1} - \Psi^1_{\cdot \wedge \sigma_1 \wedge \sigma_2} \mathbb{1}_{\cdot \leq \sigma_1} \right) \mathbb{1}_{\sigma_1 \geq \sigma_2}.
$$

We know that the term inside of the brackets is progressively measurable and that the indicator function is $\mathcal{F}_{\sigma_2}$-measurable, so progressive measurability beyond σ_2 is assured. The danger is four times less than σ_2; however, our construction ensures that for $s \leq \sigma_2$, $\Psi^1_\cdot \mathbb{1}_{\cdot \leq \sigma_1} - \Psi^1_{\cdot \wedge \sigma_1 \wedge \sigma_2} \mathbb{1}_{\cdot \leq \sigma_1} = 0$. Thus progressive measurability is established, and similarly for the final term of (3.7). The only remaining argument where $\mathcal{F}_0$-measurability was used in Proposition 3.1 was the passage through the stochastic integral in the justification of

$$
\mathbb{1}_{A_k} \int_0^{t \wedge \sigma} \mathcal{G} \Psi_s \, d\mathcal{W}_s = \mathbb{1}_{A_k} \int_0^{t \wedge \sigma^k} \mathcal{G} \Psi_s^k \, d\mathcal{W}_s,
$$

---

[3] This is why we had to ask that the process was genuinely progressively measurable, as a version may disagree at the stopped time, and hence the stopped progressively measurable process would not be a version of the stopped process itself.

equation (3.6). We take the case $k = 1$, with $k = 2$ similar. We are required to justify that

$$\mathbb{1}_{\sigma_1 \geq \sigma_2} \int_0^{t \wedge (\sigma_1 \vee \sigma_2)} \mathcal{G} \Psi_s d\mathcal{W}_s = \mathbb{1}_{\sigma_1 \geq \sigma_2} \int_0^{t \wedge \sigma^1} \mathcal{G} \Psi_s^1 d\mathcal{W}_s. \tag{3.8}$$

We immediately note that

$$\mathbb{1}_{\sigma_1 \geq \sigma_2} \int_0^{t \wedge (\sigma_1 \vee \sigma_2)} \mathcal{G} \Psi_s d\mathcal{W}_s$$

$$= \mathbb{1}_{\sigma_1 \geq \sigma_2} \int_0^{t \wedge \sigma_1} \mathcal{G} \Psi_s d\mathcal{W}_s + \mathbb{1}_{\sigma_1 \geq \sigma_2} \int_{t \wedge \sigma_1}^{t \wedge \sigma_1 \vee \sigma_2} \mathcal{G} \Psi_s d\mathcal{W}_s,$$

where we argue that the final term is null $\mathbb{P} - a.s..$ We use that the events $\{t \leq \sigma_1\}$ and $\{t > \sigma_1\}$ are $\mathcal{F}_{t \wedge \sigma_1}$-measurable and partition Ω, so

$$\int_{t \wedge \sigma_1}^{t \wedge \sigma_1 \vee \sigma_2} \mathcal{G} \Psi_s d\mathcal{W}_s$$

$$= \int_{t \wedge \sigma_1}^{t \wedge \sigma_1 \vee \sigma_2} \mathbb{1}_{t \leq \sigma_1} \mathcal{G} \Psi_s d\mathcal{W}_s + \int_{t \wedge \sigma_1}^{t \wedge \sigma_1 \vee \sigma_2} \mathbb{1}_{t > \sigma_1} \mathcal{G} \Psi_s d\mathcal{W}_s$$

$$= \int_{\sigma_1}^{t \wedge \sigma_1 \vee \sigma_2} \mathbb{1}_{t > \sigma_1} \mathcal{G} \Psi_s d\mathcal{W}_s. \tag{3.9}$$

However, as $\mathbb{1}_{\sigma_1 \geq \sigma_2}$ is $\mathcal{F}_{\sigma_1}$-measurable, then through Proposition 2.17 and the associated Remark 2.4,

$$\mathbb{1}_{\sigma_1 \geq \sigma_2} \int_{\sigma_1}^{t \wedge \sigma_1 \vee \sigma_2} \mathbb{1}_{t > \sigma_1} \mathcal{G} \Psi_s d\mathcal{W}_s = \int_{\sigma_1}^{t \wedge \sigma_1 \vee \sigma_2} \mathbb{1}_{\sigma_1 \geq \sigma_2} \mathbb{1}_{t > \sigma_1} \mathcal{G} \Psi_s d\mathcal{W}_s = 0.$$

Altogether and in light of (3.8), it is now sufficient to show

$$\mathbb{1}_{\sigma_1 \geq \sigma_2} \int_0^{t \wedge \sigma_1} \mathcal{G} \Psi_s d\mathcal{W}_s = \mathbb{1}_{\sigma_1 \geq \sigma_2} \int_0^{t \wedge \sigma_1} \mathcal{G} \Psi_s^1 d\mathcal{W}_s. \tag{3.10}$$

We use the linearity of $\mathcal{G}$ and the representation (3.7) to consider the three corresponding integrals. Firstly,

$$\mathbb{1}_{\sigma_1 \geq \sigma_2} \int_0^{t \wedge \sigma_1} \mathcal{G}(\Psi_{s \wedge \sigma_1 \wedge \sigma_2}^1) d\mathcal{W}_s$$

$$= \mathbb{1}_{\sigma_1 \geq \sigma_2} \int_0^{t \wedge \sigma_1 \wedge \sigma_2} \mathcal{G}(\Psi_{s \wedge \sigma_1 \wedge \sigma_2}^1) d\mathcal{W}_s + \mathbb{1}_{\sigma_1 \geq \sigma_2} \int_{t \wedge \sigma_1 \wedge \sigma_2}^{t \wedge \sigma_1} \mathcal{G}(\Psi_{s \wedge \sigma_1 \wedge \sigma_2}^1) d\mathcal{W}_s$$

$$= \mathbb{1}_{\sigma_1 \geq \sigma_2} \int_0^{t \wedge \sigma_1 \wedge \sigma_2} \mathcal{G}(\Psi_s^1) d\mathcal{W}_s + \mathbb{1}_{\sigma_1 \geq \sigma_2} \int_{t \wedge \sigma_1 \wedge \sigma_2}^{t \wedge \sigma_1} \mathcal{G}(\Psi_{s \wedge \sigma_1 \wedge \sigma_2}^1) d\mathcal{W}_s. \quad (3.11)$$

Secondly, recalling that the integrand is null up to σ_2 and again using linearity of $\mathcal{G}$,

$$\mathbb{1}_{\sigma_1 \geq \sigma_2} \int_0^{t \wedge \sigma_1} \mathcal{G}\left(\left(\Psi_s^1 - \Psi_{s \wedge \sigma_1 \wedge \sigma_2}^1\right) \mathbb{1}_{\sigma_1 \geq \sigma_2}\right) d\mathcal{W}_s$$

$$= \mathbb{1}_{\sigma_1 \geq \sigma_2} \int_{t \wedge \sigma_1 \wedge \sigma_2}^{t \wedge \sigma_1} \mathcal{G}\left(\left(\Psi_s^1 - \Psi_{s \wedge \sigma_1 \wedge \sigma_2}^1\right) \mathbb{1}_{\sigma_1 \geq \sigma_2}\right) d\mathcal{W}_s$$

$$= \mathbb{1}_{\sigma_1 \geq \sigma_2} \int_{t \wedge \sigma_1 \wedge \sigma_2}^{t \wedge \sigma_1} \mathcal{G}\left(\Psi_s^1\right) d\mathcal{W}_s - \mathbb{1}_{\sigma_1 \geq \sigma_2} \int_{t \wedge \sigma_1 \wedge \sigma_2}^{t \wedge \sigma_1} \mathcal{G}\left(\Psi_{s \wedge \sigma_1 \wedge \sigma_2}^1\right) d\mathcal{W}_s, \quad (3.12)$$

where we have pulled the $\mathcal{F}_{\sigma_1 \wedge \sigma_2}$-measurable $\mathbb{1}_{\sigma_1 \geq \sigma_2}$ out of the stochastic integral, eliminating t from the lower integration point with a similar treatment to (3.9). Notice that summing (3.11) and (3.12), we achieve precisely

$$\mathbb{1}_{\sigma_1 \geq \sigma_2} \int_0^{t \wedge \sigma_1} \mathcal{G}\Psi_s^1 d\mathcal{W}_s$$

as desired in (3.10). Therefore, we would conclude the proof by showing that

$$\mathbb{1}_{\sigma_1 \geq \sigma_2} \int_0^{t \wedge \sigma_1} \mathcal{G}\left(\left(\Psi_s^2 - \Psi_{s \wedge \sigma_1 \wedge \sigma_2}^2\right) \mathbb{1}_{\sigma_1 < \sigma_2}\right) d\mathcal{W}_s = 0.$$

One can readily apply the techniques used already in this proof, using cancelation of the integrand up until $\sigma_1 \wedge \sigma_2$, then carrying $\mathbb{1}_{\sigma_1 \geq \sigma_2}$ inside this integral, and obtaining complete cancelation due to the complement indicator function $\mathbb{1}_{\sigma_1 < \sigma_2}$. The proof is complete. □

Lemma 3.3 *Under the assumptions of Theorem 3.3, there exists a maximal regular solution of the equation (3.2).*

Proof We wish to show that there exists a $\Theta \in \mathcal{Y}$ such that for any $\Gamma \in \mathcal{Y}$, $\Theta \leq \Gamma$ $\mathbb{P} - a.s.$ implies $\Theta = \Gamma \mathbb{P} - a.s..$ This will be sufficient to conclude the proof, as for Θ given by the limit of (σ_j) with corresponding solutions (Ψ^j, σ_j), our process Ψ can be consistently defined on $[0, \Theta)$ through the $\mathbb{P} - a.s.$ limit of Ψ^j from the uniqueness of local regular solutions.[4] Thus, Ψ satisfies

$$\Psi(\omega) := \Psi^j(\omega) \qquad \text{on } [0, \sigma_j(\omega)]. \quad (3.13)$$

---

[4] This construction is completely analogous to the stochastic integral driven by a local martingale, see Sect. 2.2.

We apply Zorn's Lemma on $\mathscr{Y}$, which we understand to be a partially ordered set for the relation "$\leq$" defined by $\Gamma_1 \leq \Gamma_2$ if and only if for $\mathbb{P} - a.e.$ ω, $\Gamma_1(\omega) \leq \Gamma_2(\omega)$. The result would then follow from Zorn's Lemma if we can prove that for every sequence (Γ_k) in $\mathscr{Y}$ with $\Gamma_1 \leq \cdots \leq \Gamma_k \leq \Gamma_{k+1} \leq \ldots$, there exists a $\Lambda \in \mathscr{Y}$ whereby $\Gamma_k \leq \Lambda$ for all k. Suppose now that each Γ_k is given by the increasing limit of (σ_j^k). Let us define the sequence (γ_n) as

$$\gamma_n := \bigvee_{k=1}^{n} \sigma_n^k,$$

which by virtue of Lemma 3.2 is a sequence in $\mathscr{X}$. Inherited from each (σ_n^k), note that this is a $\mathbb{P} - a.s.$ monotone increasing sequence and therefore admits a limiting stopping time which we claim to be our Λ. By definition $\Lambda \in \mathscr{Y}$, and for each fixed k we see that $\gamma_n \geq \sigma_n^k$ for $n \geq k$. We thus have that the limit of the (γ_n) dominates the limit of the (σ_n^k), which proves the result. $\square$

Remark 3.3 An alternative approach could be taken without appealing to Zorn's Lemma, following [21, Section 18 pp.71]. We find our method to be simpler and more direct, hence of our preference.

Lemma 3.4 *Let* (Ψ, Θ) *be a maximal regular solution of the equation (3.2). Under the assumptions of Theorem 3.3,* (Ψ, Θ) *is unique.*

Proof We start by showing that $\Theta = \Gamma$ $\mathbb{P} - a.s.$, for Γ as in Definition 3.8. Suppose that (θ_j), (γ_j) are the sequences in $\mathscr{X}$ convergent to Θ, Γ, respectively, as stipulated in Definition 3.7. From Lemma 3.2 we have that the sequence $(\theta_j \vee \gamma_j)$ lives in $\mathscr{X}$ and is clearly monotone increasing and convergent to $\Theta \vee \Gamma$ $\mathbb{P} - a.s.$. But then $\Theta \vee \Gamma \in \mathscr{Y}$, which by the fact that $\Theta, \Gamma \leq \Theta \vee \Gamma$ $\mathbb{P} - a.s.$ and the definition of the maximal solution, $\Theta = \Theta \vee \Gamma = \Gamma$ $\mathbb{P} - a.s.$ The assumed uniqueness of the local regular solutions $((\Psi, \theta_j \vee \gamma_j))$ concludes the result. $\square$

The way we chose to define the maximal time Θ did not exclude the possibility that some $\Gamma \in \mathscr{Y}$ could be such that $\Gamma > \Theta$ on a set of positive measures, which is often excluded from the definition: See [15, 38]. The following corollary shows that such a scenario cannot occur.

Corollary 3.1 *Let* (Ψ, Θ) *be the unique maximal strong solution of the equation (3.2). Then for any* $\Gamma \in \mathscr{Y}$, $\Theta \geq \Gamma$ $\mathbb{P} - a.s..$

Proof Via the same arguments we have that $\Theta = \Theta \vee \Gamma$ $\mathbb{P} - a.s.$ which concludes the proof. $\square$

Of course, we have demonstrated everything necessary to prove Theorem 3.3.

Proof of Theorem 3.3 The proof follows from Lemmas 3.3, 3.4. $\square$

3.4   Stratonovich SPDEs in the Abstract Framework

We work now with the same initial condition Ψ_0 and operators Q, $\mathcal{G}$ but instead pose the question of how to understand the Stratonovich SPDE

$$\Psi_t = \Psi_0 + \int_0^t Q\Psi_s ds + \int_0^t \mathcal{G}\Psi_s \circ d\mathcal{W}_s. \tag{3.14}$$

We would like to avoid introducing a strong solution for the equation (3.14) as understanding the Stratonovich integral is delicate given the necessary semi-martingale structure. If we can show that Ψ satisfies an evolution equation of an Itô stochastic integral plus a time integral, then we can deduce the required semi-martingality and identify the martingale part, but we would need this assumption on semi-martingality a priori to define the Stratonovich integral in the sense of Definition 3.3 in order to show the desired representation. To this end, we identify a Stratonovich SPDE with an Itô one in the manner given here.

Theorem 3.4 *If Ψ is a strong solution of the equation*

$$\Psi_t = \Psi_0 + \int_0^t \left(Q\Psi_s + \frac{1}{2} \sum_{i=1}^{\infty} \mathcal{G}_i^2 \Psi_s \right) ds + \int_0^t \mathcal{G}\Psi_s d\mathcal{W}_s, \tag{3.15}$$

then Ψ satisfies the identity (3.14) $\mathbb{P}-a.s.$ in X for all $t \in [0, T]$.

There is a little to unpack here before going on to the proof of this result. The first is how we understand the infinite sum of (3.15) and subsequently the SPDE. The operator

$$\sum_{i=1}^{\infty} \mathcal{G}_i^2 : V \to U$$

is defined as the pointwise limit of the partial sums, which is well defined as for any fixed $\phi \in V$,

$$\left\| \sum_{i=m}^{n} \mathcal{G}_i^2 \phi \right\|_U \leq \sum_{i=m}^{n} \left\| \mathcal{G}_i^2 \phi \right\|_U \leq \sum_{i=m}^{n} c_i \left\| \mathcal{G}_i \phi \right\|_H \leq \sum_{i=m}^{n} c_i^2 \left\| \phi \right\|_V, \tag{3.16}$$

which as seen before approaches zero as $m \to \infty$. To understand the strong solution as defined in Definition 3.5, we need to show that the new operator $Q + \frac{1}{2} \sum_{i=1}^{\infty} \mathcal{G}_i^2$ satisfies the assumptions postulated in Assumption 3.1, but this is clear as $\mathcal{G}_i : V \to H$ and $\mathcal{G}_i : H \to U$ are bounded linear, hence continuous so measurable, thus too is $\mathcal{G}_i^2 : V \to U$, and therefore the partial sums and the pointwise limit are as well. Moreover

$$\left\| \sum_{i=1}^{\infty} \mathcal{G}_i^2 \phi \right\|_U \leq \sum_{i=1}^{\infty} c_i^2 \, \|\phi\|_V$$

as seen in (3.16) so the boundedness is also satisfied so we can understand the SPDE (3.15) in the same manner as (3.2). It also remains to be checked that the Stratonovich integral of (3.14) is well defined for Ψ a strong solution of (3.15). We show this in the sense of Definition 3.3 for $\mathcal{H} = X$, the space in which the identity is satisfied. We show that the sequence of stopping times

$$\tau_n := n \wedge \inf \left\{ s \geq 0 : \int_0^s \|\Psi_r\|_V^2 \, dr \geq n \right\}$$

fit the requirements of Definition 3.3, noting immediately that the sequence is $\mathbb{P} -$ a.s. monotone increasing and convergent to infinity as the process

$$s \mapsto \int_0^s \|\Psi_r\|_V^2 \, dr$$

is continuous. Moreover, the process $\mathcal{G}\Psi^n := \mathcal{G}(\Psi.\mathbb{1}_{\cdot \leq \tau_n}) = \mathcal{G}\Psi.\mathbb{1}_{\cdot \leq \tau_n}$ (using linearity of $\mathcal{G}$) is progressively measurable in H (we remark again that this is really $\mathcal{G}\Phi^n$ as stipulated in Remark 3.1, but we identify the two) and satisfies the bound

$$\mathbb{E} \int_0^t \sum_{i=1}^{\infty} \|\mathcal{G}_i \Psi_r^n\|_H^2 \, dr \leq \mathbb{E} \sum_{i=1}^{\infty} c_i^2 \int_0^t \|\Psi_r^n\|_V^2 \leq \sum_{i=1}^{\infty} c_i^2 n < \infty$$

freely applying Tonelli's Theorem between the expectation, integral, and sum. Thus $\mathcal{G}\Psi^n \in \mathcal{I}^H(\mathcal{W})$ so the integral can be constructed in H and then embedded into X, but we note that the embedding $J : H \to X$ is a continuous linear operator, and so from Proposition 2.16, then $J(\mathcal{G}\Psi^n) \in \mathcal{I}^X(\mathcal{W})$, and in particular

$$J \left(\int_0^t \mathcal{G}\Psi_s^n d\mathcal{W}_s \right) = \int_0^t J(\mathcal{G}\Psi_s^n) d\mathcal{W}_s$$

so there is no ambiguity in how we understand the integral as an element of X. Indeed, we simply make the identification $\mathcal{G}\Psi^n$ with $J(\mathcal{G}\Psi^n)$ and will make no explicit reference to the embeddings in our analysis henceforth. It remains to show that:

1. Each $\mathcal{G}_i \Psi^{\tau_n} \in \bar{\mathcal{M}}_c^2(X)$.
2. The infinite sum $\sum_{i=1}^{\infty} [\mathcal{G}_i \Psi^{\tau_n}, W^i]_t$ converges in $L^2(\Omega; X)$.

For the first point we look at the evolution equation satisfied by Ψ^{τ_n}, which is

$$\Psi_t^{\tau_n} = \Psi_0^{\tau_n} + \int_0^t \left(Q + \frac{1}{2} \sum_{i=1}^{\infty} \mathcal{G}_i^2 \right) (\Psi_s^n) \mathbb{1}_{s \leq \tau_n} ds + \int_0^t \mathcal{G}\Psi_s^n d\mathcal{W}_s$$

$\mathbb{P} - a.s.$ in U, and therefore from Proposition 2.16 we have that

$$\mathcal{G}_i \Psi_t^{\tau_n} = \mathcal{G}_i \Psi_0^{\tau_n} + \int_0^t \mathcal{G}_i \left(\left(Q + \frac{1}{2} \sum_{i=1}^{\infty} \mathcal{G}_i^2 \right) (\Psi_s^n) \mathbb{1}_{s \leq \tau_n} \right) ds + \int_0^t \mathcal{G}_i \mathcal{G} \Psi_s^n d\mathcal{W}_s$$

(3.17)

$\mathbb{P} - a.s.$ in X ($\mathcal{G}_i : U \to X$ is bounded and linear). The time integral is of bounded-variation in X, and from Proposition 2.18 we have the result. The second condition to prove is that the infinite sum

$$\sum_{i=1}^{\infty} [\mathcal{G}_i \Psi^{\tau_n}, W^i]_t$$

(3.18)

converges in $L^2(\Omega; X)$. From the identity (3.17) and the definition of the cross-variation for the semi-martingale,

$$[\mathcal{G}_i \Psi^{\tau_n}, W^i]_t = \left[\int_0^{\cdot} \mathcal{G}_i \mathcal{G} \Psi_s^n d\mathcal{W}_s, W^i \right]_t.$$

(3.19)

We can now use Lemma 2.6 and the definition of the integral as an $L^2(\Omega; X)$ limit to see that

$$\left[\int_0^{\cdot} \mathcal{G}_i \mathcal{G} \Psi_s^n d\mathcal{W}_s, W^i \right]_t = \left[\sum_{j=1}^{\infty} \int_0^{\cdot} \mathcal{G}_i \mathcal{G}_j \Psi_s^n d\mathcal{W}_s^j, W^i \right]_t$$

$$= \lim_{m \to \infty} \left[\sum_{j=1}^{m} \int_0^{\cdot} \mathcal{G}_i \mathcal{G}_j \Psi_s^n d\mathcal{W}_s^j, W^i \right]_t,$$

where the limit is taken in $L^1(\Omega, X)$, should this limit exist and satisfy the conditions of Lemma 2.6. We consider (a_k) as an orthonormal basis of X, so recalling Definition 2.18,

$$\left[\sum_{j=1}^{m} \int_0^{\cdot} \mathcal{G}_i \mathcal{G}_j \Psi_s^n d W_s^j, W^i \right]_t = \sum_{k=1}^{\infty} \left[\left\langle \sum_{j=1}^{m} \int_0^{\cdot} \mathcal{G}_i \mathcal{G}_j \Psi_s^n d W_s^j, a_k \right\rangle_X, W^i \right]_t a_k.$$

Now we can first use Theorem 2.1 to reduce this to

$$\sum_{k=1}^{\infty} \left[\sum_{j=1}^{m} \int_0^{\cdot} \langle \mathcal{G}_i \mathcal{G}_j \Psi_s^n, a_k \rangle_X d W_s^j, W^i \right]_t a_k$$

from which the classical real valued theory informs us that for $m \geq i$, due to the independence of the Brownian Motions, this is simply

$$\sum_{k=1}^{\infty} \left(\int_0^t \left\langle \mathcal{G}_i^2 \boldsymbol{\Psi}_s^n, a_k \right\rangle_X ds \right) a_k = \sum_{k=1}^{\infty} \left\langle \int_0^t \mathcal{G}_i^2 \boldsymbol{\Psi}_s^n ds, a_k \right\rangle_X a_k = \int_0^t \mathcal{G}_i^2 \boldsymbol{\Psi}_s^n ds.$$

Therefore the limit as $m \to \infty$ is well defined, and we have the representation for (3.19). The convergence of the infinite sum in $L^2(\Omega, X)$, (3.18), will follow from our standard Cauchy argument, though we have to work a little harder here. The Cauchy argument requires showing that

$$\mathbb{E} \left\| \sum_{i=m}^{k} \int_0^t \mathcal{G}_i^2 \boldsymbol{\Psi}_s^n ds \right\|_X^2 \longrightarrow 0$$

as $m, k \to \infty$ to which end we note that

$$\mathbb{E} \left\| \sum_{i=m}^{k} \int_0^t \mathcal{G}_i^2 \boldsymbol{\Psi}_s^n ds \right\|_X^2 \leq \mathbb{E} \left(\sum_{i=m}^{k} \int_0^t \left\| \mathcal{G}_i^2 \boldsymbol{\Psi}_s^n \right\|_X ds \right)^2$$

and

$$\mathbb{E} \left(\sum_{i=m}^{k} \int_0^t \left\| \mathcal{G}_i^2 \boldsymbol{\Psi}_s^n \right\|_X ds \right)^2 \leq \mathbb{E} \left(\sum_{i=m}^{k} c_i^2 \int_0^t \left\| \boldsymbol{\Psi}_s^n \right\|_H ds \right)^2$$

$$\leq \mathbb{E} \left(\sum_{i=m}^{k} c_i^2 \int_0^t c \left\| \boldsymbol{\Psi}_s^n \right\|_V ds \right)^2$$

$$\leq \left(c \sum_{i=m}^{k} c_i^2 \right)^2 \mathbb{E} \left[t \int_0^t \left\| \boldsymbol{\Psi}_s^n \right\|_V^2 ds \right]$$

$$\leq \left(c \sum_{i=m}^{k} c_i^2 \right)^2 tn,$$

where c is the constant from the embedding of $V \hookrightarrow H$. As $\sum_{i=1}^{\infty} c_i^2 < \infty$, then the Cauchy property follows, and hence the Stratonovich integral is indeed well defined.

Proof of Theorem 3.4 We must show that

$$\int_0^t \mathcal{G}\boldsymbol{\Psi}_s \circ d\mathcal{W}_s = \int_0^t \mathcal{G}\boldsymbol{\Psi}_s d\mathcal{W}_s + \frac{1}{2} \int_0^t \sum_{i=1}^{\infty} \mathcal{G}_i^2 \boldsymbol{\Psi}_s ds \qquad (3.20)$$

$\mathbb{P} - a.s.$ in X for all $t \geq 0$. It is sufficient to show that for any n, the identity

$$\int_0^{t \wedge \tau_n} \mathcal{G}\Psi_s \circ dW_s = \int_0^{t \wedge \tau_n} \mathcal{G}\Psi_s dW_s + \frac{1}{2} \int_0^{t \wedge \tau_n} \sum_{i=1}^{\infty} \mathcal{G}_i^2 \Psi_s ds \qquad (3.21)$$

holds $\mathbb{P} - a.s.$ in X for all $t \geq 0$. This can be seen as, defining the sets $A_n \subset \Omega$ to be those ω whereby (3.21) holds for all $t \geq 0$, then $\bigcap A_n$ is a set of full probabilities such that for every $\omega \in \bigcap A_n$, (3.21) holds for all n and t; in particular, for any $t \geq 0$ there is an n such that $\tau_n(\omega) \geq t$, for which (3.21) holds and is equivalent to (3.20) at this ω. We have that

$$\int_0^{t \wedge \tau_n} \mathcal{G}\Psi_s \circ dW_s = \int_0^{t \wedge \tau_n} \mathcal{G}_i \Psi_s dW_s + \frac{1}{2} \sum_{i=1}^{\infty} [\mathcal{G}_i \Psi^{\tau_n}, W^i]_t,$$

where the limit is taken in $L^2(\Omega; X)$. Of course we have just shown that

$$\sum_{i=1}^{\infty} [\mathcal{G}_i \Psi^{\tau_n}, W^i]_t = \sum_{i=1}^{\infty} \int_0^t \mathcal{G}_i^2 \Psi_s^n ds$$

is well defined in this topology, but we must show that it is equal to

$$\int_0^t \sum_{i=1}^{\infty} \mathcal{G}_i^2 \Psi_s^n ds$$

for the limit $\mathbb{P} - a.s.$ in U as it was defined at (3.16). By an application of the Dominated Convergence Theorem with dominating function

$$\sum_{i=1}^{\infty} c_i^2 \|\Psi^n\|_V,$$

we can rewrite

$$\int_0^t \sum_{i=1}^{\infty} \mathcal{G}_i^2 \Psi_s^n ds = \sum_{i=1}^{\infty} \int_0^t \mathcal{G}_i^2 \Psi_s^n ds$$

as a limit $\mathbb{P} - a.s.$ in U, and thus $\mathbb{P} - a.s.$ in X. However, the convergence in $L^2(\Omega; X)$ implies that of a subsequence $\mathbb{P} - a.s.$ in X, which agrees with the limit of the whole sequence $\mathbb{P} - a.s.$ in X, thus giving the result.   □

This theorem has been stated for the strong solution (Definition 3.5), though we note that all arguments follow in the corresponding local case by incorporating the stopping time as done in the justification that the integrals in Definition 3.4 are well defined. The result is stated below.

Corollary 3.2 *If* (Ψ, τ) *is a local strong solution of the equation (3.15), then* Ψ *satisfies the identity*

$$\Psi_t = \Psi_0 + \int_0^{t \wedge \tau} Q\Psi_s ds + \int_0^{t \wedge \tau} G\Psi_s \circ dW_s$$

$\mathbb{P} - a.s.$ *in X for all* $t \in [0, T]$.

Remark 3.4 We should emphasize why the identity (3.14) holds only in X and not in U, the space in which (3.15) is satisfied. The evolution equation (3.17) allowed us to identify the semi-martingale structure in X as is this is where the integrals are constructed. We can, however, construct the stochastic integral in (3.17) in U, which allows us to conclude that the time integral is itself an element of U (as $G_i \Psi_t^n, G_i \Psi_0^n$ are as well). Unfortunately we cannot say that this is actually an integral in U (just an integral in X which is in turn an element of U) so pertinently we cannot say that this is of bounded-variation in U, which would be necessary when considering the cross-variation $[G_i \Psi^n, W^i]_t$ in U.

We use Theorem 3.4 as a way of defining the Stratonovich SPDE, however as discussed, with a priori martingality assumptions, then the Stratonovich integral is well defined and one can prove a converse of this theorem.

Theorem 3.5 *Let* Ψ *be such that for* $\mathbb{P} - a.e.$ ω, $\Psi_{\cdot}(\omega) \in C([0, T]; H) \cap L^2([0, T]; V)$ *with* Ψ *progressively measurable in* V. *Assume in addition that* $G\Psi \in \bar{\mathcal{I}}_\circ^U(\mathcal{W})$ *and* Ψ *satisfies the identity (3.14)* $\mathbb{P} - a.s.$ *in* U *for all* $t \geq 0$. *Then* Ψ *satisfies the identity (3.15)* $\mathbb{P} - a.s.$ *in X for all* $t \in [0, T]$.

Proof By assumption, the Stratonovich integral is well defined so we can write

$$\Psi_t = \Psi_0 + \int_0^t Q\Psi_s ds + \sum_{i=1}^{\infty} \int_0^t G_i \Psi_s dW_s^i + \frac{1}{2} \sum_{i=1}^{\infty} \left[G_i \Psi, W^i \right]_t$$

for the cross-variation taken in U. We wish to write out this cross-variation process explicitly, so as in the proof of Theorem 3.4 we consider the evolution equation satisfied by $G_i \Psi$. We appreciate that

$$G_i \left(\sum_{j=1}^{\infty} \left[G_j \Psi, W^j \right]_t \right) = \sum_{j=1}^{\infty} G_i \left(\left[G_j \Psi, W^j \right]_t \right)$$

for the limit now in $L^2(\Omega; X)$, validated as G_i is bounded and linear from U into X. It is not clear if this is of finite-variation, so to proceed similarly to (3.19), we wish to show that

$$\left[\sum_{j=1}^{\infty} G_i \left(\left[G_j \Psi, W^j \right] \right), W^i \right]_t = 0$$

$\mathbb{P} - a.s.$ for any $t \geq 0$. For this we again use Lemma 2.6 to see that

$$\left[\sum_{j=1}^{\infty} \mathcal{G}_i \left(\left[\mathcal{G}_j \Psi, W^j \right] \right), W^i \right]_t = \lim_{m \to \infty} \left[\sum_{j=1}^{m} \mathcal{G}_i \left(\left[\mathcal{G}_j \Psi, W^j \right] \right), W^i \right]_t$$

for the limit in $L^1(\Omega; X)$. Each term in this sequence must be zero, though, as each $\left[\mathcal{G}_j \Psi, W^j \right]$ is of finite-variation in U so $\mathcal{G}_i \left(\left[\mathcal{G}_j \Psi, W^j \right] \right)$ is of finite-variation in X. Thus through considering the identity (3.14) in X, by the exact same process as Theorem 3.4, we prove the result.   □

3.5 Weak Solutions in the Abstract Framework

As in the study of PDEs, to expand the existence theory, we shall also consider weaker notions of solution. We do this in two ways: *analytically* weak solutions and *probabilistically* weak solutions, the second of which we shall refer to as martingale solutions. We begin by giving a definition of analytically weak solutions in the established framework. For this, we extend our assumptions slightly.

Assumption 3.6 $Q : H \to X$ *is measurable, and there exists a map* $\tilde{K} = \tilde{K}_{c,p,q}$ *such that for any* $\psi \in H$,

$$\|Q\psi\|_X \leq \tilde{K}(\phi)[1 + \|\psi\|_H^2].$$

In addition, we now ask that there exists a bilinear form $\langle \cdot, \cdot \rangle_{X \times H} : X \times H \to \mathbb{R}$ *such that for* $\phi \in U$, $\psi \in H$,

$$\langle \phi, \psi \rangle_{X \times H} = \langle \phi, \psi \rangle_U . \tag{3.22}$$

We further require that the inclusions $V \hookrightarrow H \hookrightarrow U$ *are dense. For the noise* $\mathcal{G}$ *we assume that there exists an adjoint operator* $\mathcal{G}_i^*$ *with the same boundedness properties as* $\mathcal{G}_i$ *satisfying*

$$\langle \mathcal{G}_i \phi, \psi \rangle_U = \langle \phi, \mathcal{G}_i^* \psi \rangle_U \tag{3.23}$$

for all $\phi, \psi \in H$.

Remark 3.5 A frequent setting is that X is given by the dual space H^*, and the bilinear form $\langle \cdot, \cdot \rangle_{X \times H}$ is the duality $\langle \cdot, \cdot \rangle_{H^* \times H}$ with respect to U. This is the induced Gelfand Triple.

To simplify notation we consider $\mathcal{G}^*$ as an operator on $\mathfrak{U}$ in the same way as we do for $\mathcal{G}$. With this in place we can define such a solution.

Definition 3.9 Let $\Psi_0 : \Omega \to U$ be $\mathcal{F}_0$-measurable. A process Ψ such that for $\mathbb{P}-$ a.e. ω, $\Psi.(\omega) \in C_w([0, T]; U) \cap L^2([0, T]; H)$ with $\Psi.$ progressively measurable in H, is said to be a weak solution of the equation (3.2) if the identity

$$\langle \Psi_t, \phi \rangle_U = \langle \Psi_0, \phi \rangle_U + \int_0^t \langle Q\Psi_s, \phi \rangle_{X \times H} \, ds + \int_0^t \langle \Psi_s, \mathcal{G}^*\phi \rangle_U \, d\mathcal{W}_s \qquad (3.24)$$

holds $\mathbb{P}-a.s.$ in $\mathbb{R}$ for all $\phi \in H$ and $t \in [0, T]$.

Immediately we observe that the stochastic integral is well defined through exactly the same justification as the integral for strong solutions. We also justify our use of the terminology "weak."

Proposition 3.2 *Let $\Psi_0 : \Omega \to H$ be $\mathcal{F}_0$-measurable, and suppose that Ψ is a process whereby for $\mathbb{P}-a.e.$ ω, $\Psi.(\omega) \in C([0, T]; H) \cap L^2([0, T]; V)$ with $\Psi.$ progressively measurable in V. Then Ψ is a strong solution of the equation (3.2) if and only if it is a weak solution.*

Proof We consider the two implications in turn, noting that we only need to show the equivalence of the two identities (3.2), (3.24):

$\Longrightarrow$: This direction is clear, as we simply take the inner product with ϕ in the identity (3.2) and then use Proposition 2.17. The properties (3.22), (3.23) then conclude the implication.

$\Longleftarrow$: Via the identical process in reverse, we obtain that

$$\langle \Psi_t, \phi \rangle_U = \langle \Psi_0, \phi \rangle_U + \left\langle \int_0^t Q\Psi_s ds, \phi \right\rangle_U + \left\langle \int_0^t \mathcal{G}\Psi_s d\mathcal{W}_s, \phi \right\rangle_U$$

for all $\phi \in H$. We then use the density of H in U to deduce this identity for all $\phi \in U$, from which the result follows.

$\square$

There is an important difference between the weak and strong solutions in terms of the Stratonovich Equation (3.14). Recall from the previous section that we experience a "loss of a derivative" from $\mathcal{G}_i$ in the Itô–Stratonovich conversion, in the sense that our solutions had to have a degree of regularity greater than what was needed to define them. In the weak form this operator hits the test function instead, so it is there where some additional regularity is required. We have the following analogue of Theorem 3.4.

Theorem 3.7 *If Ψ is a weak solution of the equation (3.15), then Ψ satisfies the identity*

$$\langle \Psi_t, \phi \rangle_U = \langle \Psi_0, \phi \rangle_U + \int_0^t \langle Q\Psi_s, \phi \rangle_{X \times H} \, ds + \int_0^t \langle \Psi_s, \mathcal{G}^*\phi \rangle_U \circ d\mathcal{W}_s \qquad (3.25)$$

$\mathbb{P}-a.s.$ in $\mathbb{R}$ for all $\phi \in V$ and $t \in [0, T]$.

As we did for Theorem 3.4, we first make precise the meaning of a weak solution of (3.15) in terms of the "Itô–Stratonovich Corrector." The infinite sum is again taken as a pointwise limit, where $\boldsymbol{\Psi}$ satisfies the identity

$$\langle \boldsymbol{\Psi}_t, \phi \rangle_U = \langle \boldsymbol{\Psi}_0, \phi \rangle_U + \int_0^t \langle Q\boldsymbol{\Psi}_s, \phi \rangle_{X \times H} \, ds$$

$$+ \frac{1}{2} \int_0^t \sum_{i=1}^\infty \langle \mathcal{G}_i \boldsymbol{\Psi}_s, \mathcal{G}_i^* \phi \rangle_U \, ds + \int_0^t \langle \boldsymbol{\Psi}_s, \mathcal{G}^* \phi \rangle_U \, d\mathcal{W}_s.$$

Many of the technicalities of this result were addressed in Theorem 3.4, so in the proof here we only show directly that the Itô–Stratonovich Corrector is of the right form, where now we consider τ^n as the first hitting time in H and use the same notation $\boldsymbol{\Psi}^n$.

Proof of Theorem 3.7 Through the arguments of Theorem 3.4, it is sufficient to show that

$$\left[\langle \boldsymbol{\Psi}^n, \mathcal{G}_i^* \phi \rangle_U, W^i \right]_t = \int_0^t \langle \mathcal{G}_i \boldsymbol{\Psi}_s^n, \mathcal{G}_i^* \phi \rangle_U \, ds \tag{3.26}$$

for any given $\phi \in V$. For this we consider the evolution equation satisfied by $\langle \boldsymbol{\Psi}^n, \mathcal{G}_i^* \phi \rangle_U$. Because of this, then $\mathcal{G}_i^* \phi \in H$ and can be used as a test function in the weak formulation, such that $\boldsymbol{\Psi}$ satisfies the identity

$$\langle \boldsymbol{\Psi}_t^n, \mathcal{G}_i^* \phi \rangle_U = \langle \boldsymbol{\Psi}_0^n, \mathcal{G}_i^* \phi \rangle_U + \int_0^t \langle Q\boldsymbol{\Psi}_s^n, \mathcal{G}_i^* \phi \rangle_{X \times H} \, ds$$

$$+ \frac{1}{2} \int_0^t \sum_{i=1}^\infty \langle \mathcal{G}_i \boldsymbol{\Psi}_s^n, \mathcal{G}_i^* \mathcal{G}_i^* \phi \rangle_U \, ds + \int_0^t \langle \boldsymbol{\Psi}_s^n, \mathcal{G}_i^* \mathcal{G}^* \phi \rangle_U \, d\mathcal{W}_s.$$

Of course we can rewrite $\langle \boldsymbol{\Psi}_s^n, \mathcal{G}_i^* \mathcal{G}^* \phi \rangle_U = \langle \mathcal{G}_i \boldsymbol{\Psi}_s^n, \mathcal{G}^* \phi \rangle_U$ so through the same process as in Theorem 3.4, the above identity implies (3.26) which concludes the proof.   □

In accordance with the loss of regularity established in Theorem 3.4, we cannot say in Theorem 3.7 that

$$\int_0^t \langle \mathcal{G}\boldsymbol{\Psi}_s, \phi \rangle_U \circ d\mathcal{W}_s = \left\langle \int_0^t \mathcal{G}\boldsymbol{\Psi}_s \circ d\mathcal{W}_s, \phi \right\rangle_U,$$

and hence this should be considered a "very weak" solution of the Stratonovich equation. To conclude this section we very briefly comment on the notion of a probabilistically weak solution, which we refer to as a martingale solution.

Definition 3.10 Let $\boldsymbol{\Psi}_0 : \Omega \to H$ be $\mathcal{F}_0$-measurable. If there exists a filtered probability space $\left(\tilde{\Omega}, \tilde{\mathcal{F}}, (\tilde{\mathcal{F}}_t), \tilde{\mathbb{P}}\right)$, a cylindrical Brownian Motion $\tilde{\mathcal{W}}$ over $\mathfrak{U}$ with respect to $\left(\tilde{\Omega}, \tilde{\mathcal{F}}, (\tilde{\mathcal{F}}_t), \tilde{\mathbb{P}}\right)$, an $\tilde{\mathcal{F}}_0$-measurable $\tilde{\boldsymbol{\Psi}}_0 : \tilde{\Omega} \to H$ with the same law as $\boldsymbol{\Psi}_0$, a process $\tilde{\boldsymbol{\Psi}}$ such that for $\tilde{\mathbb{P}} - a.e.\ \omega,\ \tilde{\boldsymbol{\Psi}}_{\cdot}(\omega) \in C\left([0, T]; H\right) \cap L^2\left([0, T]; V\right)$ with $\tilde{\boldsymbol{\Psi}}$ progressively measurable in V, is said to be a martingale strong solution of the equation (3.2) if the identity

$$\tilde{\boldsymbol{\Psi}}_t = \tilde{\boldsymbol{\Psi}}_0 + \int_0^t Q\tilde{\boldsymbol{\Psi}}_s ds + \int_0^t \mathcal{G}\tilde{\boldsymbol{\Psi}}_s d\mathcal{W}_s$$

holds $\tilde{\mathbb{P}} - a.s.$ in U for all $t \in [0, T]$.

3.6 Time-Dependent Operators

We did not facilitate time dependence in the operators $Q, \mathcal{G}$ as solely for the fact that if $\mathcal{G}$ was time dependent, then the conversion from Stratonovich to Itô form would be more troublesome. There is no real additional difficulty in establishing a framework for the Itô Form for time-dependent operators, so we briefly do so now. We only consider the Itô form here so there is no longer a need for the space X, thus we work with the triple

$$V \hookrightarrow H \hookrightarrow U,$$

and now the SPDE

$$\boldsymbol{\Psi}_t = \boldsymbol{\Psi}_0 + \int_0^t Q(s, \boldsymbol{\Psi}_s) ds + \int_0^t \mathcal{G}(s, \boldsymbol{\Psi}_s) d\mathcal{W}_s. \qquad (3.27)$$

We require the assumptions now as:

Assumption 3.8 *The operators* $Q : [0, T] \times V \to U$ *and* $\mathcal{G} : [0, T] \times V \to \mathcal{L}^2(\mathfrak{U}; H)$ *are measurable.*

Assumption 3.9 *There exists a* $C_{\cdot} : [0, T] \to \mathbb{R}$ *bounded, and constants* c_i *such that for every* $\phi \in V$ *and* $t \in [0, T]$,

$$\|Q(t, \boldsymbol{\phi})\|_U \leq C_t \tilde{K}(\phi) \left[1 + \|\boldsymbol{\phi}\|_V^2\right]$$

$$\|\mathcal{G}_i(t, \boldsymbol{\phi})\|_H^2 \leq C_t c_i(1 + \|\phi\|_V^2)$$

$$\sum_{i=1}^{\infty} c_i < \infty.$$

Definitions of solutions in this framework, and a justification that the integrals are well defined, then follow largely in the same way as for (3.2) so we omit the details here. What is slightly more delicate is the progressive measurability of the process $G(\cdot, \mathbf{\Phi}.)\mathbb{1}._{\leq \tau}$ in $\mathscr{L}^2(\mathfrak{U}; H)$. From the measurability of G and the progressive measurability of $\mathbf{\Phi}$, we have that for any fixed t, the mapping

$$G(\cdot, \mathbf{\Phi}.) : [0, t] \times [0, t] \times \Omega \to \mathscr{L}^2(\mathfrak{U}; H)$$

defined by

$$(s, r, \omega) \mapsto G(s, \mathbf{\Phi}_r(\omega))$$

is $\mathcal{B}([0, t]) \times \mathcal{B}([0, t]) \times \mathcal{F}_t$ measurable, and hence the mapping

$$(s, \omega) \mapsto G(s, \mathbf{\Phi}_s(\omega))$$

is $\mathcal{B}([0, t]) \times \mathcal{F}_t$ measurable as is the indicator up to the stopping time, so the product retains the required measurability.

Chapter 4
A Toolbox for Nonlinear SPDEs

In this chapter we present a range of results toward the analysis of nonlinear SPDEs. First of all we prove the well-posedness of an SPDE evolving in a finite dimensional Hilbert space but driven by a Cylindrical Brownian Motion, under traditional Lipschitz and linear growth constraints. Criteria to deduce the existence of a limit from a select sequence of processes are also established, through both tightness and Cauchy arguments, with particular novelty in the Cauchy approach. An energy equality for SPDEs under loose assumptions is also proven.

4.1 Existence and Uniqueness in Finite Dimensions

As suggested in the introduction of this chapter, our techniques in the existence theory center around taking finite dimensional approximations and using somewhat familiar theory. This scheme is referred to as a *Galerkin Approximation*, used traditionally in the analysis for highly nontrivial PDEs so offers a very reasonable first suggestion for the study of their stochastic counterparts. This approach will only work if we can quickly deduce the existence and uniqueness of solutions of the finite dimensional system, which we do in this section. It should be noted however that we still work with the Cylindrical Brownian Motion $\mathcal{W}$, over the same infinite dimensional Hilbert Space $\mathfrak{U}$; it is only the space in which the equation satisfies its identity that is assumed finite dimensional. To avoid concerns around solutions only existing until T when we use the Galerkin Approximation, we state this result globally.

Theorem 4.1 *Fix a finite dimensional Hilbert space $\mathcal{H}$. Suppose the following:*

1. For any $S > 0$, the operators $\mathscr{A} : [0, S] \times \mathcal{H} \to \mathcal{H}$ and $\mathscr{G} : [0, S] \times \mathcal{H} \to \mathscr{L}^2(\mathfrak{U}; \mathcal{H})$ are measurable.

D. Goodair, D. Crisan, *Stochastic Calculus in Infinite Dimensions and SPDEs*, SpringerBriefs in Mathematics, https://doi.org/10.1007/978-3-031-69586-5_4

2. *There exist a $C. : [0, \infty) \to \mathbb{R}$ bounded on $[0, S]$ for every S and constants c_i*
   such that for every $\phi, \psi \in \mathcal{H}$ and $t \in [0, \infty)$,

$$\|\mathscr{A}(t, \phi)\|_{\mathcal{H}}^2 \le C_t \left[1 + \|\phi\|_{\mathcal{H}}^2 \right]$$

$$\|\mathscr{G}_i(t, \phi)\|_{\mathcal{H}}^2 \le C_t c_i \left[1 + \|\phi\|_{\mathcal{H}}^2 \right]$$

$$\sum_{i=1}^{\infty} c_i < \infty$$

$$\|\mathscr{A}(t, \phi) - \mathscr{A}(t, \psi)\|_{\mathcal{H}}^2 + \sum_{i=1}^{\infty} \|\mathscr{G}_i(t, \phi) - \mathscr{G}_i(t, \psi)\|_{\mathcal{H}}^2 \le C_t \|\phi - \psi\|_{\mathcal{H}}^2.$$

3. $\Phi_0 \in L^2(\Omega; \mathcal{H})$.

Then there exists a process $\Phi : [0, \infty) \times \Omega \to \mathcal{H}$ such that for $\mathbb{P} - a.e.$ ω, $\Phi.(\omega) \in$
$C([0, S]; \mathcal{H})$ for every $S > 0$, Φ is progressively measurable in $\mathcal{H}$, and the identity

$$\Phi_t = \Phi_0 + \int_0^t \mathscr{A}(s, \Phi_s)ds + \int_0^t \mathscr{G}(s, \Phi_s)d\mathcal{W}_s \tag{4.1}$$

holds $\mathbb{P} - a.s.$ in $\mathcal{H}$ for every $t \ge 0$.

We remark that the operators satisfy the assumptions of 3.8, 3.9 for the spaces $V = H = U := \mathcal{H}$ and that the conclusion of this theorem is the existence of a strong solution of (4.1) in the sense of Definition 3.5, taken beyond T.

Proof With $\mathcal{H}$ finite dimensional, we first restrict ourselves to finitely many Brownian Motions in our stochastic integral to make things classical. Fixing any $S > 0$, let us define the operator $\mathscr{G}^k : [0, S] \times \mathcal{H} \times \mathfrak{U} \to \mathcal{H}$ on the basis vectors (e_i) of $\mathfrak{U}$ by $\mathscr{G}^k(\cdot, \cdot, e_i) = \mathscr{G}(\cdot, \cdot, e_i)$ for $i \le k$, and zero otherwise. We consider at first the equation

$$\Phi_t^k = \Phi_0^k + \int_0^t \mathscr{A}(s, \Phi_s^k)ds + \int_0^t \mathscr{G}^k(s, \Phi_s^k)d\mathcal{W}_s$$

or equivalently

$$\Phi_t^k = \Phi_0^k + \int_0^t \mathscr{A}(s, \Phi_s^k)ds + \sum_{i=1}^{k} \int_0^t \mathscr{G}_i(s, \Phi_s^k)dW_s^i$$

for $\Phi_0^k := \Phi_0$. The existence and uniqueness of solutions to this finite dimensional system is then classical (for solutions defined as in the theorem), see, e.g., [44, Theorems 5.2.5 and 5.2.9]. Consider now solutions Φ^j, Φ^k for $j < k$ arbitrary, which therefore satisfy the difference equation

$$\Phi_r^k - \Phi_r^j = \int_0^r \mathscr{A}(s, \Phi_s^k) - \mathscr{A}(s, \Phi_s^j)ds + \int_0^r \mathscr{G}^k(s, \Phi_s^k) - \mathscr{G}^j(s, \Phi_s^j)dW_s$$

for any $r \in [0, \infty)$. By applying the energy identity Proposition 4.2, for the spaces all taken to be $\mathcal{H}$, we see further that the identity

$$\left\| \Phi_r^k - \Phi_r^j \right\|_{\mathcal{H}}^2 = 2 \int_0^r \left\langle \mathscr{A}(s, \Phi_s^k) - \mathscr{A}(s, \Phi_s^j), \Phi_s^k - \Phi_s^j \right\rangle_{\mathcal{H}} ds$$

$$+ \int_0^r \left\| \mathscr{G}^k(s, \Phi_s^k) - \mathscr{G}^j(s, \Phi_s^j) \right\|_{\mathscr{L}^2(\mathfrak{U};\mathcal{H})}^2 ds$$

$$+ 2 \int_0^r \left\langle \mathscr{G}^k(s, \Phi_s^k) - \mathscr{G}^j(s, \Phi_s^j), \Phi_s^k - \Phi_s^j \right\rangle_{\mathcal{H}} dW_s$$

holds $\mathbb{P} - a.s.$. We use Cauchy–Schwarz to move to an inequality and rewrite the quadratic variation term to give us the bound

$$\left\| \Phi_r^k - \Phi_r^j \right\|_{\mathcal{H}}^2$$

$$\leq \int_0^r \left(2 \left\| \mathscr{A}(s, \Phi_s^k) - \mathscr{A}(s, \Phi_s^j) \right\|_{\mathcal{H}} \left\| \Phi_s^k - \Phi_s^j \right\|_{\mathcal{H}} + \sum_{i=1}^j \left\| \mathscr{G}_i(s, \Phi_s^k) - \mathscr{G}_i(s, \Phi_s^j) \right\|^2 \right) ds$$

$$+ \int_0^r \sum_{i=j+1}^k \left\| \mathscr{G}_i(s, \Phi_s^k) \right\|_{\mathcal{H}}^2 ds + 2 \int_0^r \left\langle \mathscr{G}^k(s, \Phi_s^k) - \mathscr{G}^j(s, \Phi_s^j), \Phi_s^k - \Phi_s^j \right\rangle_{\mathcal{H}} dW_s.$$

In one step now we bound the stochastic integral by its absolute value, take the supremum over all such r up to any arbitrary time $t \in [0, \infty)$, and employ the Lipschitz assumption to see that

$$\sup_{r \in [0,t]} \left\| \Phi_r^k - \Phi_r^j \right\|_{\mathcal{H}}^2 \leq c \int_0^t \left\| \Phi_s^k - \Phi_s^j \right\|_{\mathcal{H}}^2 ds + \int_0^t \sum_{i=j+1}^k \left\| \mathscr{G}_i(s, \Phi_s^k) \right\|_{\mathcal{H}}^2 ds$$

$$+ 2 \sup_{r \in [0,t]} \left| \int_0^r \left\langle \mathscr{G}^k(s, \Phi_s^k) - \mathscr{G}^j(s, \Phi_s^j), \Phi_s^k - \Phi_s^j \right\rangle_{\mathcal{H}} dW_s \right|$$

for a generic constant c, allowed to depend on t. We want to take the expectation here but have to be slightly careful in ensuring that the expectation is finite; to this end we consider the stopping times

$$\tau_R := R \wedge \inf\{s \geq 0 : \max\{ \left\| \Phi_s^k \right\|_{\mathcal{H}}, \left\| \Phi_s^j \right\|_{\mathcal{H}} \} \geq R\}$$

and the process defined for any fixed R by

$$\tilde{\Phi}_s^k := \Phi_s^k \mathbb{1}_{s \le \tau_R}, \qquad \tilde{\Phi}_s^j := \Phi_s^j \mathbb{1}_{s \le \tau_R}.$$

From the continuity of the processes Φ^k, Φ^j, then (τ_R) is a $\mathbb{P} - a.s.$ monotone increasing sequence convergent to infinity and $\max\{\left\|\tilde{\Phi}_s^k\right\|_{\mathcal{H}}, \left\|\tilde{\Phi}_s^j\right\|_{\mathcal{H}}\} \le R$ for any $s \ge 0$. It is trivial that these processes satisfy the same inequality

$$\sup_{r \in [0,t]} \left\|\tilde{\Phi}_r^k - \tilde{\Phi}_r^j\right\|_{\mathcal{H}}^2 \le c \int_0^t \left\|\tilde{\Phi}_s^k - \tilde{\Phi}_s^j\right\|_{\mathcal{H}}^2 ds + \int_0^t \sum_{i=j+1}^k \left\|\mathscr{G}_i(s, \tilde{\Phi}_s^k)\right\|_{\mathcal{H}}^2 ds$$

$$+ 2 \sup_{r \in [0,t]} \left| \int_0^r \left\langle \mathscr{G}^k(s, \tilde{\Phi}_s^k) - \mathscr{G}^j(s, \tilde{\Phi}_s^j), \tilde{\Phi}_s^k - \tilde{\Phi}_s^j \right\rangle_{\mathcal{H}} dW_s \right|$$

and justify that the expectation of all terms involved is finite (indeed for the stochastic integral, using the Lipschitz assumption and the boundedness of $\tilde{\Phi}_.^k - \tilde{\Phi}_.^j$ then $\left\langle \mathscr{G}^k(\cdot, \tilde{\Phi}_.^k) - \mathscr{G}^j(\cdot, \tilde{\Phi}_.^j), \tilde{\Phi}_.^k - \tilde{\Phi}_.^j \right\rangle_{\mathcal{H}} \in I^{\mathbb{R}}(\mathcal{W})$ so the expectation of this term is finite). We do now take the expectation and apply the classical Burkholder–Davis–Gundy Inequality, Theorem A.5, to give us the bound

$$\mathbb{E} \sup_{r \in [0,t]} \left\|\tilde{\Phi}_r^k - \tilde{\Phi}_r^j\right\|_{\mathcal{H}}^2 \le c\mathbb{E} \int_0^t \left\|\tilde{\Phi}_s^k - \tilde{\Phi}_s^j\right\|_{\mathcal{H}}^2 ds + \mathbb{E} \int_0^t \sum_{i=j+1}^k \left\|\mathscr{G}_i(s, \tilde{\Phi}_s^k)\right\|_{\mathcal{H}}^2 ds$$

$$+ c\mathbb{E} \left(\int_0^t \sum_{i=1}^\infty \left\langle \mathscr{G}_i^k(s, \tilde{\Phi}_s^k) - \mathscr{G}_i^j(s, \tilde{\Phi}_s^j), \tilde{\Phi}_s^k - \tilde{\Phi}_s^j \right\rangle_{\mathcal{H}}^2 ds \right)^{\frac{1}{2}},$$

which we promptly reduce to

$$\mathbb{E} \sup_{r \in [0,t]} \left\|\tilde{\Phi}_r^k - \tilde{\Phi}_r^j\right\|_{\mathcal{H}}^2 \le c\mathbb{E} \int_0^t \left\|\tilde{\Phi}_s^k - \tilde{\Phi}_s^j\right\|_{\mathcal{H}}^2 ds + \mathbb{E} \int_0^t \sum_{i=j+1}^k \left\|\mathscr{G}_i(s, \tilde{\Phi}_s^k)\right\|_{\mathcal{H}}^2 ds$$

$$+ c\mathbb{E} \left(\int_0^t \left[\sum_{i=1}^j \left\|\mathscr{G}_i(s, \tilde{\Phi}_s^k) - \mathscr{G}_i(s, \tilde{\Phi}_s^j)\right\|^2 + \sum_{i=j+1}^k \left\|\mathscr{G}_i(s, \tilde{\Phi}_s^k)\right\|_{\mathcal{H}}^2 \right] \left\|\tilde{\Phi}_s^k - \tilde{\Phi}_s^j\right\|_{\mathcal{H}}^2 ds \right)^{\frac{1}{2}}.$$

Employing the Lipschitz assumption once more, followed by an application of Young's Inequality, we have that

$$c \left(\int_0^t \left[\sum_{i=1}^j \left\|\mathscr{G}_i(s, \tilde{\Phi}_s^k) - \mathscr{G}_i(s, \tilde{\Phi}_s^j)\right\|^2 + \sum_{i=j+1}^k \left\|\mathscr{G}_i(s, \tilde{\Phi}_s^k)\right\|_{\mathcal{H}}^2 \right] \left\|\tilde{\Phi}_s^k - \tilde{\Phi}_s^j\right\|_{\mathcal{H}}^2 ds \right)^{\frac{1}{2}}$$

$$\leq c \left(\int_0^t \left[\left\| \tilde{\Phi}_s^k - \tilde{\Phi}_s^j \right\|_{\mathcal{H}}^2 + \sum_{i=j+1}^k \left\| \mathcal{G}_i(s, \tilde{\Phi}_s^k) \right\|_{\mathcal{H}}^2 \right] \left\| \tilde{\Phi}_s^k - \tilde{\Phi}_s^j \right\|_{\mathcal{H}}^2 ds \right)^{\frac{1}{2}}$$

$$\leq c \left(\sup_{r \in [0,t]} \left\| \tilde{\Phi}_r^k - \tilde{\Phi}_r^j \right\|_{\mathcal{H}}^2 \int_0^t \left\| \tilde{\Phi}_s^k - \tilde{\Phi}_s^j \right\|_{\mathcal{H}}^2 + \sum_{i=j+1}^k \left\| \mathcal{G}_i(s, \tilde{\Phi}_s^k) \right\|_{\mathcal{H}}^2 ds \right)^{\frac{1}{2}}$$

$$= c \left(\sup_{r \in [0,t]} \left\| \tilde{\Phi}_r^k - \tilde{\Phi}_r^j \right\|_{\mathcal{H}}^2 \right)^{\frac{1}{2}} \left(\int_0^t \left\| \tilde{\Phi}_s^k - \tilde{\Phi}_s^j \right\|_{\mathcal{H}}^2 + \sum_{i=j+1}^k \left\| \mathcal{G}_i(s, \tilde{\Phi}_s^k) \right\|_{\mathcal{H}}^2 ds \right)^{\frac{1}{2}}$$

$$\leq \frac{1}{2} \sup_{r \in [0,t]} \left\| \tilde{\Phi}_r^k - \tilde{\Phi}_r^j \right\|_{\mathcal{H}}^2 + \frac{c^2}{2} \int_0^t \left(\left\| \tilde{\Phi}_s^k - \tilde{\Phi}_s^j \right\|_{\mathcal{H}}^2 + \sum_{i=j+1}^k \left\| \mathcal{G}_i(s, \tilde{\Phi}_s^k) \right\|_{\mathcal{H}}^2 \right) ds,$$

and furthermore

$$\mathbb{E} \sup_{r \in [0,t]} \left\| \tilde{\Phi}_r^k - \tilde{\Phi}_r^j \right\|_{\mathcal{H}}^2 \leq c \mathbb{E} \int_0^t \left\| \tilde{\Phi}_s^k - \tilde{\Phi}_s^j \right\|_{\mathcal{H}}^2 ds + c \mathbb{E} \int_0^t \sum_{i=j+1}^k \left\| \mathcal{G}_i(s, \tilde{\Phi}_s^k) \right\|_{\mathcal{H}}^2 ds.$$

$$(4.2)$$

It is then a standard application of the Grönwall inequality that

$$\mathbb{E} \sup_{r \in [0,t]} \left\| \tilde{\Phi}_r^k - \tilde{\Phi}_r^j \right\|_{\mathcal{H}}^2 \leq c \mathbb{E} \int_0^t \sum_{i=j+1}^k \left\| \mathcal{G}_i(s, \tilde{\Phi}_s^k) \right\|_{\mathcal{H}}^2 ds, \qquad (4.3)$$

where the new constant c incorporates e^{ct} for the constant in (4.2). Observe also that, through very similar arguments just using the linear growth property instead of the Lipschitz one, we have that

$$\mathbb{E} \sup_{r \in [0,t]} \left\| \tilde{\Phi}_r^k \right\|_{\mathcal{H}}^2 \leq \mathbb{E} \left\| \Phi_0 \right\|_{\mathcal{H}}^2$$

$$+ \mathbb{E} \int_0^t \left(2 \left\| \mathcal{A}(s, \tilde{\Phi}_s^k) \right\|_{\mathcal{H}} \left\| \tilde{\Phi}_s^k \right\|_{\mathcal{H}} + \sum_{i=1}^k \left\| \mathcal{G}_i(s, \tilde{\Phi}_s^k) \right\|_{\mathcal{H}}^2 \right) ds$$

$$+ 2 \mathbb{E} \sup_{r \in [0,t]} \left| \sum_{i=1}^k \int_0^r \left\langle \mathcal{G}_i(s, \tilde{\Phi}_s^k), \tilde{\Phi}_s^k \right\rangle_{\mathcal{H}} dW_s^i \right|$$

$$\leq \mathbb{E} \left\| \tilde{\Phi}_0 \right\|_{\mathcal{H}}^2 + c\mathbb{E} \int_0^t (1 + \left\| \tilde{\Phi}_s^k \right\|_{\mathcal{H}}) \left\| \tilde{\Phi}_s^k \right\|_{\mathcal{H}} + 1 + \left\| \tilde{\Phi}_s^k \right\|_{\mathcal{H}}^2 ds$$

$$+ c\mathbb{E} \left(\int_0^t \left(1 + \left\| \tilde{\Phi}_s^k \right\|_{\mathcal{H}}^2 \right) \left\| \tilde{\Phi}_s^k \right\|_{\mathcal{H}}^2 ds \right)^{\frac{1}{2}}$$

to which we apply $\left\| \tilde{\Phi}_s^k \right\|_{\mathcal{H}} \leq 1 + \left\| \tilde{\Phi}_s^k \right\|_{\mathcal{H}}^2$ to see that

$$\mathbb{E} \sup_{r \in [0,t]} \left\| \tilde{\Phi}_r^k \right\|_{\mathcal{H}}^2 \leq \mathbb{E} \left\| \tilde{\Phi}_0 \right\|_{\mathcal{H}}^2 + c\mathbb{E} \int_0^t 1 + \left\| \tilde{\Phi}_s^k \right\|_{\mathcal{H}}^2 ds$$

$$+ c\mathbb{E} \left(\int_0^t \left(1 + \left\| \tilde{\Phi}_s^k \right\|_{\mathcal{H}}^2 \right) \left\| \tilde{\Phi}_s^k \right\|_{\mathcal{H}}^2 ds \right)^{\frac{1}{2}}$$

and further

$$\mathbb{E} \sup_{r \in [0,t]} \left\| \tilde{\Phi}_r^k \right\|_{\mathcal{H}}^2 \leq c \left[\mathbb{E} \left\| \tilde{\Phi}_0 \right\|_{\mathcal{H}}^2 + 1 \right] + c\mathbb{E} \int_0^t \left\| \tilde{\Phi}_s^k \right\|_{\mathcal{H}}^2 ds$$

as above, integrating the 1 and adding it as a constant. Thus we have

$$\mathbb{E} \sup_{r \in [0,t]} \left\| \tilde{\Phi}_r^k \right\|_{\mathcal{H}}^2 \leq c \left[\mathbb{E} \left\| \tilde{\Phi}_0 \right\|_{\mathcal{H}}^2 + 1 \right] = c \left[\mathbb{E} \| \Phi_0 \|_{\mathcal{H}}^2 + 1 \right],$$

which is a bound uniform in k and independent of τ_R. Moreover for each fixed k we appreciate that the sequence of random variables

$$\sup_{r \in [0,t]} \left\| \tilde{\Phi}_r^k \right\|_{\mathcal{H}}^2$$

is monotone increasing (indexed by R) and convergent to $\sup_{r \in [0,t]} \left\| \Phi_r^k \right\|_{\mathcal{H}}^2$, $\mathbb{P} -$ a.s.. Thus we may apply the Monotone Convergence Theorem to this sequence of random variables to see that

$$\mathbb{E} \sup_{r \in [0,t]} \left\| \Phi_r^k \right\|_{\mathcal{H}}^2 = \lim_{R \to \infty} \mathbb{E} \sup_{r \in [0,t]} \left\| \tilde{\Phi}_r^k \right\|_{\mathcal{H}}^2 \leq c \left[\mathbb{E} \| \Phi_0 \|_{\mathcal{H}}^2 + 1 \right].$$

With this bound established we can revert back to (4.3), combining with the boundedness of the $\mathscr{G}_i$ to deduce that

$$\mathbb{E} \int_0^t \sum_{i=j+1}^k \left\| \mathscr{G}_i(s, \Phi_s^k) \right\|_{\mathcal{H}}^2 ds < \infty$$

and clearly

$$\mathbb{E}\int_0^t \sum_{i=j+1}^k \left\|\mathscr{G}_i(s,\tilde{\Phi}_s^k)\right\|_{\mathcal{H}}^2 ds \le \mathbb{E}\int_0^t \sum_{i=j+1}^k \left\|\mathscr{G}_i(s,\Phi_s^k)\right\|_{\mathcal{H}}^2 ds.$$

So we can update (4.3) with the bound

$$\mathbb{E}\sup_{r\in[0,t]}\left\|\tilde{\Phi}_r^k - \tilde{\Phi}_r^j\right\|_{\mathcal{H}}^2 \le c\mathbb{E}\int_0^t \sum_{i=j+1}^k \left\|\mathscr{G}_i(s,\Phi_s^k)\right\|_{\mathcal{H}}^2 ds$$

to which we apply the same monotone convergence argument to deduce that

$$\mathbb{E}\sup_{r\in[0,t]}\left\|\Phi_r^k - \Phi_r^j\right\|_{\mathcal{H}}^2 \le c\mathbb{E}\int_0^t \sum_{i=j+1}^k \left\|\mathscr{G}_i(s,\Phi_s^k)\right\|_{\mathcal{H}}^2 ds. \tag{4.4}$$

Moreover

$$\sup_{k>j}\mathbb{E}\int_0^t \sum_{i=j+1}^k \left\|\mathscr{G}_i(s,\Phi_s^k)\right\|_{\mathcal{H}}^2 ds \le t\sup_{k>j}\mathbb{E}\sup_{r\in[0,t]}\sum_{i=j+1}^k \left\|\mathscr{G}_i(r,\Phi_r^k)\right\|_{\mathcal{H}}^2$$

$$\le t\sup_{k>j}\mathbb{E}\sup_{r\in[0,t]}\sum_{i=j+1}^k C_t c_i\left(1 + \left\|\Phi_r^k\right\|_{\mathcal{H}}^2\right)$$

$$\le c\sup_{k>j}\sum_{i=j+1}^k c_i\mathbb{E}\sup_{r\in[0,t]}\left(1 + \left\|\Phi_r^k\right\|_{\mathcal{H}}^2\right)$$

$$\le c\sum_{i=j+1}^\infty c_i\sup_{k>j}\mathbb{E}\sup_{r\in[0,t]}\left(1 + \left\|\Phi_r^k\right\|_{\mathcal{H}}^2\right)$$

$$\le c\sum_{i=j+1}^\infty c_i\left(c\left[\mathbb{E}\|\Phi_0\|_{\mathcal{H}}^2 + 1\right]\right),$$

which is a sequence in j monotone decreasing to zero. Note that we have absorbed a time dependence into the constant c, which is not meaningful. Thus in view of (4.4),

$$\lim_{j\to\infty}\sup_{k>j}\left[\mathbb{E}\sup_{r\in[0,t]}\left\|\Phi_r^k - \Phi_r^j\right\|_{\mathcal{H}}^2\right] = 0,$$

so the sequence (Φ^k) is Cauchy in $L^2(\Omega; C([0, t]; \mathcal{H}))$ and as such we can deduce the existence of a Φ such that $\Phi^k \rightarrow \Phi$ in this space (and hence, in $L^2(\Omega; L^2([0, t]; \mathcal{H}))$) for every $t \in [0, S]$, and thus Φ is also the $\mathbb{P} - a.s.$ limit of a subsequence of the (Φ^k) in $C([0, S]; \mathcal{H})$. This limit process inherits the progressive measurability (indeed it is adapted and has continuous paths in $\mathcal{H}$). It simply remains to show that Φ satisfies the identity (4.1), so we first consider the $\mathbb{P} - a.s.$ convergent subsequence (Φ^{k_l}). Looking at the stochastic integral and employing Proposition 2.15, we have that

$$
\mathbb{E} \left\| \int_0^t \mathscr{G}(s, \Phi_s) d\mathcal{W}_s - \int_0^t \mathscr{G}^{k_l}(s, \Phi_s^{k_l}) d\mathcal{W}_s \right\|_{\mathcal{H}}^2
$$

$$
= \mathbb{E} \left\| \int_0^t \mathscr{G}(s, \Phi_s) - \mathscr{G}^{k_l}(s, \Phi_s^{k_l}) d\mathcal{W}_s \right\|_{\mathcal{H}}^2
$$

$$
= \mathbb{E} \int_0^t \sum_{i=1}^{\infty} \left\| \mathscr{G}_i(s, \Phi_s) - \mathscr{G}_i^{k_l}(s, \Phi_s^{k_l}) \right\|_{\mathcal{H}}^2 ds
$$

$$
= \mathbb{E} \int_0^t \left(\sum_{i=1}^{k_l} \left\| \mathscr{G}_i(s, \Phi_s) - \mathscr{G}_i(s, \Phi_s^{k_l}) \right\|_{\mathcal{H}}^2 ds + \sum_{i=k_l+1}^{\infty} \| \mathscr{G}_i(s, \Phi_s) \|_{\mathcal{H}}^2 \right) ds
$$

$$
\leq \sum_{i=1}^{k_l} \mathbb{E} \int_0^t c_i \left\| \Phi_s - \Phi_s^{k_l} \right\|_{\mathcal{H}}^2 ds + \sum_{i=k_l+1}^{\infty} \mathbb{E} \int_0^t c_i \| \Phi_s \|_{\mathcal{H}}^2 ds
$$

$$
\leq c\mathbb{E} \int_0^t \left\| \Phi_s - \Phi_s^{k_l} \right\|_{\mathcal{H}}^2 ds + \sum_{i=k_l+1}^{\infty} \mathbb{E} \int_0^t c_i \| \Phi_s \|_{\mathcal{H}}^2 ds.
$$

So from the known $L^2(\Omega; L^2([0, t]; \mathcal{H}))$ convergence, we have that

$$
\lim_{k_l \to \infty} \int_0^t \mathscr{G}^{k_l}(s, \Phi_s^{k_l}) d\mathcal{W}_s = \int_0^t \mathscr{G}(s, \Phi_s) d\mathcal{W}_s
$$

with the limit taken in $L^2(\Omega; \mathcal{H})$. We can thus extract a further subsequence which we denote (Φ^{k_m}) such that this limit holds $\mathbb{P} - a.s.$ in $\mathcal{H}$ and is of course still such that $(\Phi^{k_m}) \rightarrow \Phi \ \mathbb{P} - a.s.$ in $L^2([0, t]; \mathcal{H})$. Therefore

$$
\left\| \int_0^t \mathscr{A}(s, \Phi_s) ds - \int_0^t \mathscr{A}(s, \Phi_s^{k_m}) ds \right\|_{\mathcal{H}}^2 \leq t \int_0^t \left\| \mathscr{A}(s, \Phi_s) - \mathscr{A}(s, \Phi_s^{k_m}) \right\|_{\mathcal{H}}^2 ds
$$

$$
\leq ct \int_0^t \left\| \Phi_s - \Phi_s^{k_m} \right\|_{\mathcal{H}}^2 ds
$$

and so

$$\lim_{k_m \to \infty} \int_0^t \mathscr{A}(s, \Phi_s^{k_m}) ds = \int_0^t \mathscr{A}(s, \Phi_s) ds$$

with the limit $\mathbb{P} - a.s.$ in $\mathcal{H}$. Thus by taking the $\mathbb{P} - a.s.$ limit in $\mathcal{H}$ of the identity satisfied by Φ^{k_m}, we reach (4.1) as required. □

Theorem 4.2 *Suppose* Ψ *is another strong solution of (4.1). Then for every* $S \geq 0$,

$$\mathbb{P}(\{\omega \in \Omega : \Phi_t(\omega) = \Psi_t(\omega) \quad \forall t \in [0, S]\}) = 1.$$

Proof The method of proof here is entirely contained in that for the existence proof. Indeed we look at the energy equality satisfied by the difference of the solutions, which is

$$\|\Phi_r - \Psi_r\|_{\mathcal{H}}^2 = 2 \int_0^r \langle \mathscr{A}(s, \Phi_s) - \mathscr{A}(s, \Psi_s), \Phi_s - \Psi_s \rangle_{\mathcal{H}} ds$$

$$+ \int_0^r \|\mathscr{G}(s, \Phi_s) - \mathscr{G}(s, \Psi_s)\|_{\mathscr{L}^2(\mathfrak{U};\mathcal{H})}^2 ds$$

$$+ 2 \int_0^r \langle \mathscr{G}(s, \Phi_s) - \mathscr{G}(s, \Psi_s), \Phi_s - \Psi_s \rangle_{\mathcal{H}} d\mathcal{W}_s.$$

Following along the proof, we introduce

$$\tau_R := R \wedge \inf\{s \geq 0 : \max\{\|\Phi_s\|_{\mathcal{H}}, \|\Psi_s\|_{\mathcal{H}}\} \geq R\}$$

and the process defined for any fixed R by

$$\tilde{\Phi}_s := \Phi_s \mathbb{1}_{s \leq \tau_R}, \qquad \tilde{\Psi}_s := \Psi_s \mathbb{1}_{s \leq \tau_R}.$$

In this case we have the inequality

$$\sup_{r \in [0,t]} \left\| \tilde{\Phi}_r - \tilde{\Psi}_r \right\|_{\mathcal{H}}^2 \leq c \int_0^t \left\| \tilde{\Phi}_s - \tilde{\Psi}_s \right\|_{\mathcal{H}}^2 ds$$

$$+ 2 \sup_{r \in [0,t]} \left| \int_0^r \left\langle \mathscr{G}(s, \tilde{\Phi}_s) - \mathscr{G}(s, \tilde{\Psi}_s), \tilde{\Phi}_s - \tilde{\Psi}_s \right\rangle_{\mathcal{H}} d\mathcal{W}_s \right|$$

so following all of the same steps, simply now without the $\sum_{i=j+1}^k \left\| \mathscr{G}_i(s, \tilde{\Phi}_s^k) \right\|_{\mathcal{H}}^2$ term, we deduce again that

$$\mathbb{E} \sup_{r \in [0,t]} \left\| \tilde{\Phi}_r - \tilde{\Psi}_r \right\|_{\mathcal{H}}^2 \leq 0$$

in analogy with (4.3). By the same monotone convergence argument, we have that

$$\mathbb{E} \sup_{r \in [0,t]} \| \boldsymbol{\Phi}_r - \boldsymbol{\Psi}_r \|_{\mathcal{H}}^2 = 0,$$

which gives the result.   □

4.2   Tightness Criteria

Unsurprisingly, one route into the relative compactness methods of PDE theory is through tightness, owing to Prokhorov's Theorem. Poignantly we can connect this weak limit of measures with a genuine process through Skorohod's Representation Theorem, see, for example, [7, pp. 70]. Simple criteria through which we can establish tightness in the space of solutions to SPDEs will prove useful. Recalling our notions of solution, for example, Definitions 3.5 and 3.9, we consider tightness criteria in both the spaces $L^2 ([0, T]; \mathcal{H})$ and $\mathcal{D} ([0, T]; \mathcal{H})$ for suitably chosen $\mathcal{H}$ (recall the definition of $\mathcal{D} ([0, T]; \mathcal{H})$ from Sect. 1.2). We note, for example, [7, pp. 124], that the Skorohod Topology is equivalent to the uniform topology when restricted to continuous functions. It is necessary to use due to the separability of the associated metric space, allowing us to invoke Prokhorov's Theorem. This process is reviewed at the end of the section.

We first give two results for tightness in $\mathcal{D} ([0, T]; \mathcal{H})$. This method was applied in [59] but was not established into a result, so we give a full proof here.

Lemma 4.1 *Let $\mathcal{Y}$ be a reflexive Banach Space and $\mathcal{H}$ a Hilbert Space such that $\mathcal{Y}$ is compactly embedded into $\mathcal{H}$, and consider the induced Gelfand Triple*

$$\mathcal{Y} \hookrightarrow \mathcal{H} \hookrightarrow \mathcal{Y}^*.$$

For some fixed $T > 0$ let $\boldsymbol{\Psi}^n : \Omega \to C ([0, T]; \mathcal{H})$ be a sequence of measurable processes such that for every $t \in [0, T]$,

$$\sup_{n \in \mathbb{N}} \mathbb{E} \left(\sup_{t \in [0,T]} \| \boldsymbol{\Psi}_t^n \|_{\mathcal{H}} \right) < \infty \qquad (4.5)$$

and for any sequence of stopping times (γ_n) with $\gamma_n : \Omega \to [0, T]$, and any $\varepsilon > 0$, $y \in \mathcal{Y}$,

$$\lim_{\delta \to 0^+} \sup_{n \in \mathbb{N}} \mathbb{P} \left(\left\{ \omega \in \Omega : \left| \left\langle \boldsymbol{\Psi}_{(\gamma_n + \delta) \wedge T}^n - \boldsymbol{\Psi}_{\gamma_n}^n, y \right\rangle_{\mathcal{H}} \right| > \varepsilon \right\} \right) = 0. \qquad (4.6)$$

Then the sequence of the laws of $(\boldsymbol{\Psi}^n)$ is tight in the space of probability measures over $\mathcal{D} ([0, T]; \mathcal{Y}^*)$.

Proof We essentially combine the tightness criteria of Theorems A.6 and A.7, in the specific case outlined here. First in reference to Theorem A.7 we may take E to be $\mathcal{Y}^*$ and $\mathbb{F}$ to be $(\mathcal{Y}^*)^*$, which is well known to separate points in $\mathcal{Y}^*$ from a corollary of the Hahn–Banach Theorem which asserts that for every $\phi \in \mathcal{Y}^*$ there exists a $\psi \in (\mathcal{Y}^*)^*$ such that $\langle \phi, \psi \rangle_{\mathcal{Y}^* \times (\mathcal{Y}^*)^*} = \|\phi\|_{\mathcal{Y}^*}$. We also note that condition 2 in Theorem A.7 is satisfied for (μ_n) taken to be the sequence of laws of (Ψ^n) over $\mathcal{D}([0, T]; \mathcal{Y}^*)$, owing to the property (4.5). Indeed as $\mathcal{Y}$ is compactly embedded into $\mathcal{H}$, then $\mathcal{H}$ is compactly embedded into $\mathcal{Y}^*$, so one only needs to take a bounded subset of $\mathcal{H}$ for this condition. Considering the closed ball of radius M in $\mathcal{H}$, $\tilde{B}_M$, we have that

$$\mathbb{P}\left(\left\{\omega \in \Omega : \Psi^n(\omega) \notin D\left([0, T]; \tilde{B}_M\right)\right\}\right) \le \mathbb{P}\left(\left\{\omega \in \Omega : \Psi^n(\omega) \notin C\left([0, T]; \tilde{B}_M\right)\right\}\right)$$

$$\le \mathbb{P}\left(\left\{\omega \in \Omega : \sup_{t \in [0,T]} \left\|\Psi^n_t(\omega)\right\|_{\mathcal{H}} > M\right\}\right)$$

$$\le \frac{1}{M}\mathbb{E}\left(\sup_{t \in [0,T]} \left\|\Psi^n_t\right\|_{\mathcal{H}}\right)$$

$$\le \frac{1}{M}\sup_{n \in \mathbb{N}} \mathbb{E}\left(\sup_{t \in [0,T]} \left\|\Psi^n_t\right\|_{\mathcal{H}}\right)$$

from which we see an arbitrarily large choice of M will justify condition 2 of Theorem A.7. By applying the theorem, it only remains to show that for every $\psi \in (\mathcal{Y}^*)^*$ the sequence of the laws of $\langle \Psi^n, \psi \rangle_{\mathcal{Y}^* \times (\mathcal{Y}^*)^*}$ is tight in the space of probability measures over $\mathcal{D}([0, T]; \mathbb{R})$. By the reflexivity of $\mathcal{Y}$ for every $\psi \in (\mathcal{Y}^*)^*$, there exists a $y \in \mathcal{Y}$ such that $\langle \Psi^n, \psi \rangle_{\mathcal{Y}^* \times (\mathcal{Y}^*)^*} = \langle \Psi^n, y \rangle_{\mathcal{Y}^* \times \mathcal{Y}}$, and as $\Psi^n_t \in \mathcal{H} \, \mathbb{P} - a.s.$, then this is furthermore just $\langle \Psi^n, y \rangle_{\mathcal{H}}$. The problem is now reduced to showing tightness in $\mathcal{D}([0, T]; \mathbb{R})$, which by Theorem A.6 is satisfied if we can show that for any sequence of stopping times (γ_n), $\gamma_n : \Omega \to [0, T]$, and constants (δ_n), $\delta_n \ge 0$, and $\delta_n \to 0$ as $n \to \infty$:

1. For every $t \in [0, T]$, the sequence of the laws of $\langle \Psi^n_t, y \rangle_{\mathcal{H}}$ is tight in the space of probability measures over $\mathbb{R}$.
2. For every $\varepsilon > 0$, $\lim_{n \to \infty} \mathbb{P}\left(\left\{\omega \in \Omega : \left|\langle \Psi^n_{(\gamma_n + \delta_n) \wedge T} - \Psi^n_{\gamma_n}, y \rangle_{\mathcal{H}}\right| > \varepsilon\right\}\right) = 0$.

We address each item in turn: As for 1, we are required to show that for every $\varepsilon > 0$ and $t \in [0, T]$, there exists a compact $K_\varepsilon \subset \mathbb{R}$ such that for every $n \in \mathbb{N}$,

$$\mathbb{P}\left(\left\{\omega \in \Omega : \langle \Psi^n_t(\omega), y \rangle_{\mathcal{H}} \notin K_\varepsilon\right\}\right) < \varepsilon.$$

To this end define B_M as the closed ball of radius M in $\mathbb{R}$, and then

$$\mathbb{P}\left(\left\{\omega \in \Omega : \langle \Psi^n_t(\omega), y \rangle_{\mathcal{H}} \notin B_M\right\}\right) = \mathbb{P}\left(\left\{\omega \in \Omega : \left|\langle \Psi^n_t(\omega), y \rangle_{\mathcal{H}}\right| > M\right\}\right)$$

$$\leq \frac{1}{M} \mathbb{E}(|\langle \boldsymbol{\Psi}_t^n, y \rangle_{\mathcal{H}}|)$$

$$\leq \frac{\|y\|_{\mathcal{H}}}{M} \sup_{n \in \mathbb{N}} \mathbb{E}\left(\|\boldsymbol{\Psi}_t^n\|_{\mathcal{H}}\right)$$

so setting

$$M := \frac{2\|y\|_{\mathcal{H}} \sup_{n \in \mathbb{N}} \mathbb{E}\left(\|\boldsymbol{\Psi}_t^n\|_{\mathcal{H}}\right)}{\varepsilon}$$

justifies item 1. As for 2, note that for each fixed $j \in \mathbb{N}$ we have that

$$\left|\left\langle \boldsymbol{\Psi}_{(\gamma_j + \delta_j) \wedge T}^j - \boldsymbol{\Psi}_{\gamma_j}^j, y \right\rangle_{\mathcal{H}}\right| \leq \sup_{n \in \mathbb{N}} \left|\left\langle \boldsymbol{\Psi}_{(\gamma_n + \delta_j) \wedge T}^n - \boldsymbol{\Psi}_{\gamma_n}^n, y \right\rangle_{\mathcal{H}}\right|$$

so in particular

$$\lim_{j \to \infty} \mathbb{P}\left(\left\{\left|\left\langle \boldsymbol{\Psi}_{(\gamma_j + \delta_j) \wedge T}^j - \boldsymbol{\Psi}_{\gamma_j}^j, y \right\rangle_{\mathcal{H}}\right| > \varepsilon\right\}\right)$$

$$\leq \lim_{j \to \infty} \sup_{n \in \mathbb{N}} \mathbb{P}\left(\left\{\left|\left\langle \boldsymbol{\Psi}_{(\gamma_n + \delta_j) \wedge T}^n - \boldsymbol{\Psi}_{\gamma_n}^n, y \right\rangle_{\mathcal{H}}\right| > \varepsilon\right\}\right).$$

As (δ_j) was an arbitrary sequence of nonnegative constants approaching zero, we can generically take $\delta \to 0^+$ and 2 is implied by (4.6). The proof is complete. $\square$

We do not need to rely on the Gelfand Triple structure to obtain such a criterion.

Lemma 4.2 *Let $\mathcal{H}_1, \mathcal{H}_2$ be Hilbert spaces with $\mathcal{H}_1$ compactly embedded into $\mathcal{H}_2$, and V any dense set in $\mathcal{H}_2$. For some fixed $T > 0$, let $\boldsymbol{\Psi}^n : \Omega \to C\left([0, T]; \mathcal{H}_1\right)$ be a sequence of measurable processes such that*

$$\sup_{n \in \mathbb{N}} \mathbb{E}\left(\sup_{t \in [0,T]} \|\boldsymbol{\Psi}_t^n\|_{\mathcal{H}_1}\right) < \infty \qquad (4.7)$$

and for any sequence of stopping times (γ_n) with $\gamma_n : \Omega \to [0, T]$, and any $\varepsilon > 0$, $v \in V$,

$$\lim_{\delta \to 0^+} \sup_{n \in \mathbb{N}} \mathbb{P}\left(\left\{\omega \in \Omega : \left|\left\langle \boldsymbol{\Psi}_{(\gamma_n + \delta) \wedge T}^n - \boldsymbol{\Psi}_{\gamma_n}^n, v \right\rangle_{\mathcal{H}_2}\right| > \varepsilon\right\}\right) = 0. \qquad (4.8)$$

Then the sequence of the laws of $(\boldsymbol{\Psi}^n)$ is tight in the space of probability measures over $\mathcal{D}\left([0, T]; \mathcal{H}_2\right)$.

Proof The proof is mechanically nearly identical to that of Lemma 4.1, so we highlight only the slight technical differences. In reference to Theorem A.7 we may take E to be $\mathcal{H}_2$ and $\mathbb{F}$ to be the collection of functions defined for each

$v \in V$ by $\langle \cdot, v \rangle_{\mathcal{H}_2}$, which separates points in $\mathcal{H}_2$ from the density of V. We also note that condition 2 of Theorem A.7 is satisfied for (μ_n) taken to be the sequence of laws of $(\mathbf{\Psi}^n)$ over $\mathcal{D}([0, T]; \mathcal{H}_2)$, owing to the property (4.7). Indeed as $\mathcal{H}_1$ is compactly embedded into $\mathcal{H}_2$, one only needs to take a bounded subset of $\mathcal{H}_1$; hence considering the closed ball of radius M in $\mathcal{H}_1$, $\tilde{B}_M$, we have that

$$\mathbb{P}\left(\left\{\omega \in \Omega : \mathbf{\Psi}^n(\omega) \notin \mathcal{D}\left([0, T]; \tilde{B}_M\right)\right\}\right) \leq \frac{1}{M} \sup_{n \in \mathbb{N}} \mathbb{E}(\sup_{t \in [0,T]} \|\mathbf{\Psi}_t^n\|_{\mathcal{H}_1})$$

precisely as in Lemma 4.1, from which we see an arbitrarily large choice of M will justify (3.3). Therefore by applying Theorem A.7 it only remains to show that for every $v \in V$ the sequence of the laws of $\langle \mathbf{\Psi}^n, v \rangle_{\mathcal{H}_2}$ is tight in the space of probability measures over $\mathcal{D}([0, T]; \mathbb{R})$. The remainder of the proof now follows exactly as in Lemma 4.1.   □

For completeness, we state a criterion from the literature for tightness in the $L^2([0, T]; \mathcal{H})$ space as well. A brief proof is given here, which is taken from [59, Lemma 5.2].

Lemma 4.3 *Let $\mathcal{H}_1, \mathcal{H}_2$ be Hilbert Spaces such that $\mathcal{H}_1$ is compactly embedded into $\mathcal{H}_2$, and for some fixed $T > 0$ let $(\mathbf{\Psi}^n) : \Omega \times [0, T] \to \mathcal{H}_1$ be a sequence of measurable processes such that*

$$\sup_{n \in \mathbb{N}} \mathbb{E} \int_0^T \|\mathbf{\Psi}_s^n\|_{\mathcal{H}_1}^2 ds < \infty, \tag{4.9}$$

and for any $\varepsilon > 0$,

$$\lim_{\delta \to 0^+} \sup_{n \in \mathbb{N}} \mathbb{P}\left(\left\{\omega \in \Omega : \int_0^{T-\delta} \|\mathbf{\Psi}_{s+\delta}^n(\omega) - \mathbf{\Psi}_s^n(\omega)\|_{\mathcal{H}_2}^2 ds > \varepsilon\right\}\right) = 0. \tag{4.10}$$

Then the sequence of the laws of $(\mathbf{\Psi}^n)$ is tight in the space of probability measures over $L^2([0, T]; \mathcal{H}_2)$.

Proof Observe that from Chebyshev's inequality and condition (4.9),

$$\lim_{R \to \infty} \sup_{n \in \mathbb{N}} \mathbb{P}\left(\left\{\omega \in \Omega : \int_0^T \|\mathbf{\Psi}_s^n\|_{\mathcal{H}_1}^2 ds > R\right\}\right)$$

$$\leq \lim_{R \to \infty} \frac{1}{R}\left(\sup_{n \in \mathbb{N}} \mathbb{E} \int_0^T \|\mathbf{\Psi}_s^n\|_{\mathcal{H}_1}^2 ds\right) = 0$$

so in particular for any given $\varepsilon > 0$, there exists an R such that

$$\sup_{n \in \mathbb{N}} \mathbb{P}\left(\left\{\omega \in \Omega : \int_0^T \|\mathbf{\Psi}_s^n\|_{\mathcal{H}_1}^2 ds > R\right\}\right) < \frac{\varepsilon}{2}. \tag{4.11}$$

Define B_R to be the closed ball of radius R in $L^2([0, T]; \mathcal{H}_2)$. From (4.10) for any $k \in \mathbb{N}$ there exists a $\delta_k > 0$ such that

$$\sup_{n \in \mathbb{N}} \mathbb{P}\left(\left\{\omega \in \Omega : \int_0^{T-\delta_k} \left\|\Psi_{s+\delta_k}^n(\omega) - \Psi_s^n(\omega)\right\|_{\mathcal{H}_2}^2 ds > \frac{1}{k}\right\}\right) \le \frac{\varepsilon}{2^{k+1}}. \quad (4.12)$$

Defining

$$\Gamma_k := \left\{\phi \in L^2([0, T]; \mathcal{H}_2) : \int_0^{T-\delta_k} \left\|\phi_{s+\delta_k}(\omega) - \phi_s(\omega)\right\|_{\mathcal{H}_2}^2 ds \le \frac{1}{k}\right\}$$

it follows that the set

$$A := B_R \cap \bigcap_{k=1}^{\infty} \Gamma_k$$

is shown to be relatively compact in $L^2([0, T]; \mathcal{H}_2)$, and

$$\sup_{n \in \mathbb{N}} \mathbb{P}\left(\Psi^n \notin A\right) \le \sup_{n \in \mathbb{N}} \mathbb{P}\left(\Psi^n \notin B_R\right) + \sum_{k=1}^{\infty} \sup_{n \in \mathbb{N}} \mathbb{P}\left(\Psi^n \notin \Gamma_k\right) \le \frac{\varepsilon}{2} + \sum_{k=1}^{\infty} \frac{\varepsilon}{2^{k+1}} \le \varepsilon,$$

which concludes the proof.   □

For clarity, we lay out some of the steps that one may take to use this theory in order to deduce the existence of solutions. We refer to strong solutions of the equation (3.2), defined in Definition 3.4:

- Consider a Galerkin Approximation with solutions (Ψ^n) existing at least locally in V, as described in Sect. 4.1 and applying Theorem 4.1.
- Use the tightness criteria to show that the (Ψ^n) are tight in the spaces $\mathcal{D}([0, T]; H)$ and $L^2([0, T]; V)$, hence in their intersection endowed with the sum metric, which is again a complete separable metric space.
- Apply Prokhorov's Theorem to deduce that the sequence of the laws of (Ψ^n) is relatively compact in the space of probability measures over $\mathcal{D}([0, T]; H) \cap L^2([0, T]; V)$ endowed with the topology of weak convergence. We remark again that the space $\mathcal{D}([0, T]; H)$ is separable, while $C([0, T]; H)$ is not, hence why we have appealed to the Skorohod Space.
- Apply Skorohod's Representation Theorem to deduce the existence of a new probability space, a sequence of $\mathcal{D}([0, T]; H) \cap L^2([0, T]; V)$ valued random variables on this new probability space with the same laws as a subsequence of the (Ψ^n), and a random variable $\tilde{\Psi}$ admitted as an almost sure limit on this space. As the restriction of the $\mathcal{D}([0, T]; H)$ topology to continuous functions is equivalent to the uniform topology, then this limit $\tilde{\Psi}$ is continuous in H.

- Use the powerful almost sure convergence to deduce that $\tilde{\Psi}$ is a strong solution of the equation (3.2) posed on the new probability space; this is a *probabilistically weak* or *martingale* solution.
- Prove the uniqueness of solutions on this space, and apply a Yamada–Watanabe type argument (e.g., [58]) to obtain the existence of a strong solution on the original probability space.

Of course, when dealing with highly nontrivial SPDEs, the approach has to be fine-tuned with potential truncations and appeals to alternate spaces. This is simply a sketch of the rough idea; one can see a full, concrete approach in the related paper [36].

4.3 Cauchy Criteria

A more direct way to deduce the existence of a limiting process from the Galerkin Approximations comes from the Cauchy Property. In practice, due to potential nonlinearities, one requires some truncation to get sufficient control on the approximating sequence. More precisely for the nth term of the sequence, one must work up to a first hitting time of this process, giving a stopping time τ_n. The question is then whether we can deduce a limiting process up to some time τ, where $0 < \tau \le \tau_n$ for all n. Such a result was proven by Glatt-Holtz and Ziane in [32, Lemma 5.1].

The result presented here is an extension of this and has important applications in the deduction of *maximal* and *global* solutions. The result of Glatt-Holtz and Ziane asserts that, under assumptions of a Cauchy property of the sequence of processes up until their first hitting times and some weak equicontinuity at *time zero*, then a limiting process and positive stopping time exist (which are then argued to be a local strong solution, as a limit of the Galerkin Approximation). No characterization of this stopping time is given though; hence completely separate arguments are required to consider what interval the solution exists upon. We demonstrate that if instead one imposes that the processes satisfy a weak equicontinuity assumption at *all* times, then the limiting stopping time can be taken as a first hitting time of the limiting process for an arbitrarily large hitting parameter. Application of this result *immediately* yields that solutions exist up until they blow up, removing the need for further analysis toward the interval on which solutions exist.

Proposition 4.1 *Fix $T > 0$. For $t \in [0, T]$ let X_t denote a Banach space with norm $\|\cdot\|_{X,t}$ such that for all $s > t$, $X_s \hookrightarrow X_t$ and $\|\cdot\|_{X,t} \le \|\cdot\|_{X,s}$. Suppose that (Ψ^n) is a sequence of processes $\Psi^n : \Omega \mapsto X_T$, $\|\Psi^n\|_{X,\cdot}$ is adapted and $\mathbb{P} - a.s.$ continuous, $\Psi^n \in L^2(\Omega; X_T)$, and such that $\sup_n \|\Psi^n\|_{X,0} \in L^\infty(\Omega; \mathbb{R})$. For any given $M > 1$, define the stopping times*

$$\tau_n^{M,T} := T \wedge \inf\left\{s \ge 0 : \|\Psi^n\|_{X,s}^2 \ge M + \|\Psi^n\|_{X,0}^2\right\}. \tag{4.13}$$

Furthermore, suppose

$$\lim_{m\to\infty} \sup_{n\geq m} \mathbb{E}\left[\left\|\boldsymbol{\Psi}^n - \boldsymbol{\Psi}^m\right\|^2_{X,\tau_m^{M,T}\wedge\tau_n^{M,T}}\right] = 0 \tag{4.14}$$

and that for any stopping time γ and a sequence of stopping times (δ_j) that converge to $0\ \mathbb{P} - a.s.$,

$$\lim_{j\to\infty} \sup_{n\in\mathbb{N}} \mathbb{E}\left(\left\|\boldsymbol{\Psi}^n\right\|^2_{X,(\gamma+\delta_j)\wedge\tau_n^{M,T}} - \left\|\boldsymbol{\Psi}^n\right\|^2_{X,\gamma\wedge\tau_n^{M,T}}\right) = 0. \tag{4.15}$$

Then there exists a stopping time $\tau_\infty^{M,T}$, a process $\boldsymbol{\Psi} : \Omega \mapsto X_{\tau_\infty^{M,T}}$ whereby $\|\boldsymbol{\Psi}\|_{X,\cdot\wedge\tau_\infty^{M,T}}$ is adapted and $\mathbb{P} - a.s.$ continuous, and a subsequence indexed by (m_j) such that:

- $\tau_\infty^{M,T} \leq \tau_{m_j}^{M,T}\ \mathbb{P} - a.s..$
- $\lim_{j\to\infty} \|\boldsymbol{\Psi} - \boldsymbol{\Psi}^{m_j}\|_{X,\tau_\infty^{M,T}} = 0\ \mathbb{P} - a.s..$

Moreover for any $R > 0$ we can choose M to be such that the stopping time

$$\tau^{R,T} := T \wedge \inf\left\{s \geq 0 : \|\boldsymbol{\Psi}\|^2_{X,s\wedge\tau_\infty^{M,T}} \geq R\right\} \tag{4.16}$$

satisfies $\tau^{R,T} \leq \tau_\infty^{M,T}\ \mathbb{P} - a.s..$ Thus $\tau^{R,T}$ is simply $T \wedge \inf\left\{s \geq 0 : \|\boldsymbol{\Psi}\|^2_{X,s} \geq R\right\}$.

Remark 4.1 A consequence of the properties that $\sup_n \|\boldsymbol{\Psi}^n\|_{X,0} \in L^\infty(\Omega; \mathbb{R})$ and $\lim_{j\to\infty} \|\boldsymbol{\Psi} - \boldsymbol{\Psi}^{m_j}\|_{X,\tau_\infty^{M,T}} = 0\ \mathbb{P} - a.s.$ is that $\|\boldsymbol{\Psi}\|_{X,0} \in L^\infty(\Omega; \mathbb{R})$. Therefore for $R > \left\|\|\boldsymbol{\Psi}\|_{X,0}\right\|_{L^\infty(\Omega;\mathbb{R})}$, we have that $\tau^{R,T}$ is $\mathbb{P} - a.s.$ positive; hence so too is $\tau_\infty^{M,T}$ for appropriately chosen M.

Proof Property (4.14) implies that for any given $j \in \mathbb{N}$ we can choose an $n_j \in \mathbb{N}$ such that for all $k \geq n_j$,

$$\mathbb{E}\left(\left\|\boldsymbol{\Psi}^k - \boldsymbol{\Psi}^{n_j}\right\|^2_{X,\tau_{n_j}^{M,t}\wedge\tau_k^{M,t}}\right) \leq 2^{-4j}. \tag{4.17}$$

We shall make use of delicate manipulations of the subsequence indexed by (n_j), and for this we introduce a new sequence of stopping times. We now impose that

$$M > 2 + \left\|\sup_{n\in\mathbb{N}} \|\boldsymbol{\Psi}^n\|^2_{X,0}\right\|_{L^\infty(\Omega;\mathbb{R})}$$

and define

$$\tilde{M}^2 := \frac{M - \left\| \sup_{n \in \mathbb{N}} \| \boldsymbol{\Psi}^n \|_{X,0}^2 \right\|_{L^\infty(\Omega;\mathbb{R})}}{2} > 1.$$

The purpose of this is to define

$$\sigma_j^M := T \wedge \inf \left\{ s > 0 : \| \boldsymbol{\Psi}^{n_j} \|_{X,s} \geq (\tilde{M} - 1 + 2^{-j}) + \| \boldsymbol{\Psi}^{n_j} \|_{X,0} \right\}$$

and ensure that $\sigma_j^M \leq \tau_{n_j}^{M,T}$ at every ω. Note the key difference in not squaring the norm, and also that $\tilde{M} > 1$ so each σ_j^M is necessarily positive. To demonstrate the inequality, it is sufficient to show that, $\mathbb{P} - a.s.$,

$$\left((\tilde{M} - 1 + 2^{-j}) + \| \boldsymbol{\Psi}^{n_j} \|_{X,0} \right)^2 \leq M + \| \boldsymbol{\Psi}^{n_j} \|_{X,0}^2, \tag{4.18}$$

or more easily

$$\left(\tilde{M} + \| \boldsymbol{\Psi}^{n_j} \|_{X,0} \right)^2 \leq M + \| \boldsymbol{\Psi}^{n_j} \|_{X,0}^2.$$

This is possible as

$$\left(\tilde{M} + \| \boldsymbol{\Psi}^{n_j} \|_{X,0} \right)^2 \leq 2\tilde{M}^2 + 2 \| \boldsymbol{\Psi}^{n_j} \|_{X,0}^2 = M - \left\| \sup_n \| \boldsymbol{\Psi}^n \|_{X,0}^2 \right\|_{L^\infty(\Omega;\mathbb{R})} + 2 \| \boldsymbol{\Psi}^{n_j} \|_{X,0}^2$$

$$\leq M + \| \boldsymbol{\Psi}^{n_j} \|_{X,0}^2.$$

The property (4.18) is thus verified, so $\sigma_j^M \leq \tau_{n_j}^{M,T}$, and hence, the subsequence $(\boldsymbol{\Psi}^{n_j})$ enjoys the same properties up until the corresponding σ_j^M. In particular from (4.17),

$$\mathbb{E} \left(\| \boldsymbol{\Psi}^{n_{j+1}} - \boldsymbol{\Psi}^{n_j} \|_{X,\sigma_j^M \wedge \sigma_{j+1}^M} \right) \leq \left[\mathbb{E} \left(\| \boldsymbol{\Psi}^{n_{j+1}} - \boldsymbol{\Psi}^{n_j} \|_{X,\sigma_j^M \wedge \sigma_{j+1}^M}^2 \right) \right]^{\frac{1}{2}} \leq 2^{-2j}, \tag{4.19}$$

and hence in defining the sets

$$\Omega_j := \left\{ \omega \in \Omega : \| \boldsymbol{\Psi}^{n_{j+1}}(\omega) - \boldsymbol{\Psi}^{n_j}(\omega) \|_{X,\sigma_j^M(\omega) \wedge \sigma_{j+1}^M(\omega)} < 2^{-(j+2)} \right\}, \tag{4.20}$$

we have, by Chebyshev's Inequality and (4.19),

$$\mathbb{P} \left(\Omega_j^C \right) \leq 2^{j+2} \mathbb{E} \left(\| \boldsymbol{\Psi}^{n_{j+1}} - \boldsymbol{\Psi}^{n_j} \|_{X,\sigma_j^M \wedge \sigma_{j+1}^M} \right) \leq 2^{-j+2}.$$

We have, therefore, that

$$\sum_{j=1}^{\infty} \mathbb{P}\left(\Omega_j^C\right) < \infty$$

from which we see

$$\mathbb{P}\left(\bigcap_{K=1}^{\infty} \bigcup_{j=K}^{\infty} \Omega_j^C\right) = 0$$

courtesy of the Borel–Cantelli Lemma. It then follows that the set

$$\hat{\Omega} := \bigcup_{K=1}^{\infty} \bigcap_{j=K}^{\infty} \Omega_j$$

is such that $\mathbb{P}(\hat{\Omega}) = 1$ so that in verifying $\mathbb{P} - a.s.$ properties, we can in fact simply show that they hold *everywhere* on $\hat{\Omega}$. More precisely, we also take $\hat{\Omega}$ to be such that every $\|\Psi^{n_j}\|_{X,\cdot}$ is continuous on $\hat{\Omega}$, which is only a further countable intersection of full measure sets. We proceed by considering the sets

$$\hat{\Omega}_K := \bigcap_{j=K}^{\infty} \Omega_j$$

with the idea to just show such properties on $\hat{\Omega}_K$ for all K (as their union makes up $\hat{\Omega}$). We look to construct a new stopping time σ_∞^M (which will prove to be the desired $\tau_\infty^{M,T}$) given as the $\mathbb{P} - a.e.$ limit of (σ_j^M), built from demonstrating that (σ_j^M) is monotone decreasing everywhere on $\hat{\Omega}_K$ for all K. In other words we show that for sufficiently large j (in fact, just $j \geq K$) that the set

$$\Gamma := \{\sigma_j^M < \sigma_{j+1}^M\} \cap \hat{\Omega}_K \tag{4.21}$$

is empty. First we observe from the strict inequality $\sigma_j^M < \sigma_{j+1}^M$ on this set that $\sigma_j^M < T$, implying that

$$\sigma_j^M = \inf\left\{s > 0 : \|\Psi^{n_j}\|_{X,s} \geq (\tilde{M} - 1 + 2^{-j}) + \|\Psi^{n_j}\|_{X,0}\right\}$$

so by the continuity of $\|\Psi^{n_j}\|_{X,\cdot}$,

$$\|\Psi^{n_j}\|_{X,\sigma_j^M} = (\tilde{M} - 1 + 2^{-j}) + \|\Psi^{n_j}\|_{X,0}. \tag{4.22}$$

Using the definition of Ω_j, (4.20), for $j \geq K$, we have that

$$\left\| \boldsymbol{\Psi}^{n_j} \right\|_{X, \sigma_j^M \wedge \sigma_{j+1}^M} - \left\| \boldsymbol{\Psi}^{n_{j+1}} \right\|_{X, \sigma_j^M \wedge \sigma_{j+1}^M} \leq \left\| \boldsymbol{\Psi}^{n_{j+1}} - \boldsymbol{\Psi}^{n_j} \right\|_{X, \sigma_j^M \wedge \sigma_{j+1}^M} < 2^{-(j+2)}$$
(4.23)

and also

$$\left\| \boldsymbol{\Psi}^{n_{j+1}} \right\|_{X,0} - \left\| \boldsymbol{\Psi}^{n_j} \right\|_{X,0} \leq \left\| \boldsymbol{\Psi}^{n_{j+1}} - \boldsymbol{\Psi}^{n_j} \right\|_{X,0} < 2^{-(j+2)}.$$
(4.24)

Combining (4.22), (4.23), and (4.24), while using that $\sigma_j^M < \sigma_{j+1}^M$, we see that

$$
\begin{aligned}
\left\| \boldsymbol{\Psi}^{n_{j+1}} \right\|_{X, \sigma_j^M \wedge \sigma_{j+1}^M} &> \left\| \boldsymbol{\Psi}^{n_j} \right\|_{X, \sigma_j^M \wedge \sigma_{j+1}^M} - 2^{-(j+2)} \\
&= \left\| \boldsymbol{\Psi}^{n_j} \right\|_{X, \sigma_j^M} - 2^{-(j+2)} \\
&= \left\| \boldsymbol{\Psi}^{n_j} \right\|_{X,0} + (\tilde{M} - 1 + 2^{-j}) - 2^{-(j+2)} \\
&> \left\| \boldsymbol{\Psi}^{n_{j+1}} \right\|_{X,0} - 2^{-(j+2)} + (\tilde{M} - 1 + 2^{-j}) - 2^{-(j+2)} \\
&= \left\| \boldsymbol{\Psi}^{n_{j+1}} \right\|_{X,0} + (\tilde{M} - 1 + 2^{-(j+1)}),
\end{aligned}
$$
(4.25)

where in the last line we have used

$$- 2^{-(j+2)} - 2^{-(j+2)} + 2^{-j} = 2^{-(j+1)}.$$

The hard work is done in showing that the set Γ defined in (4.21) is empty, as on this set note that

$$\left\| \boldsymbol{\Psi}^{n_{j+1}} \right\|_{X, \sigma_j^M \wedge \sigma_{j+1}^M} \leq \left\| \boldsymbol{\Psi}^{n_{j+1}} \right\|_{X, \sigma_{j+1}^M} \leq \left\| \boldsymbol{\Psi}^{n_{j+1}} \right\|_{X,0} + (\tilde{M} - 1 + 2^{-(j+1)}),$$

which contradicts (4.25), hence Γ must be empty. Thus on every $\hat{\Omega}_K$, and furthermore the whole of $\hat{\Omega}$, the sequence (σ_j^M) is eventually monotone decreasing (and bounded below by 0). Furthermore we define σ_∞^M as the pointwise limit $\lim_{j \to \infty} \sigma_j^M$ on $\hat{\Omega}$, which must itself be a stopping time as the $\mathbb{P} - a.s.$ limit of stopping times. As mentioned, this shall prove to be our $\tau_\infty^{M,T}$, and for the existence of $\boldsymbol{\Psi}$, we show that on $\hat{\Omega}$ the subsequence $(\boldsymbol{\Psi}^{n_j})$ is Cauchy in $X_{\sigma_\infty^M}$. Every $\omega \in \hat{\Omega}$ belongs to $\hat{\Omega}_K$ for some K, and furthermore to $\hat{\Omega}_L$ for all $L > K$. We fix arbitrary $\omega \in \hat{\Omega}$ and select an associated K. At this ω, for any $j > k \geq K$, observe that

$$
\begin{aligned}
\left\| \boldsymbol{\Psi}^{n_j} - \boldsymbol{\Psi}^{n_k} \right\|_{X, \sigma_\infty^M} &= \left\| \boldsymbol{\Psi}^{n_j} - \boldsymbol{\Psi}^{n_{k+1}} + \boldsymbol{\Psi}^{n_{k+1}} - \boldsymbol{\Psi}^{n_k} \right\|_{X, \sigma_\infty^M} \\
&\leq \left\| \boldsymbol{\Psi}^{n_j} - \boldsymbol{\Psi}^{n_{k+1}} \right\|_{X, \sigma_\infty^M} + \left\| \boldsymbol{\Psi}^{n_{k+1}} - \boldsymbol{\Psi}^{n_k} \right\|_{X, \sigma_\infty^M} \\
&\leq \left\| \boldsymbol{\Psi}^{n_j} - \boldsymbol{\Psi}^{n_{k+1}} \right\|_{X, \sigma_\infty^M} + 2^{-(k+2)}
\end{aligned}
$$

$$\leq \sum_{l=k}^{j} 2^{-(l+2)}$$

$$\leq 2^{-(k+1)}$$

having carried out an inductive argument in the penultimate step. We are thus free to take K large enough so that this difference is arbitrarily small; therefore there exists a limit in the Banach Space $X_{\sigma_\infty^M}$, which we call $\mathbf{\Psi}$. The process $\|\mathbf{\Psi}\|_{X,\cdot\wedge\sigma_\infty^M}$ is adapted and $\mathbb{P} - a.s.$ continuous, as

$$\sup_{r\in[0,T]}\left|\|\mathbf{\Psi}\|_{X,r\wedge\sigma_\infty^M} - \|\mathbf{\Psi}^{n_j}\|_{X,r\wedge\sigma_\infty^M}\right| \leq \sup_{r\in[0,T]}\left|\|\mathbf{\Psi} - \mathbf{\Psi}^{n_j}\|_{X,r\wedge\sigma_\infty^M}\right|$$

$$= \|\mathbf{\Psi} - \mathbf{\Psi}^{n_j}\|_{X,\sigma_\infty^M},$$

which has $\mathbb{P} - a.s.$ limit as $j \to \infty$ equal to zero. Thus $\|\mathbf{\Psi}\|_{X,\cdot\wedge\sigma_\infty^M}$ is given, $\mathbb{P}-a.s.$, as the uniform in time limit of adapted and continuous processes, verifying the result. Moving on, it is now that we make use of (4.15) much in the same way as we did for (4.14). This will be done in the context of $\gamma := \sigma_\infty^M$ and $\delta_j := \sigma_j^M - \sigma_\infty^M$. Indeed for any $j \in \mathbb{N}$ we can choose an $m_j \in \mathbb{N}$ (where $m_j = n_l$ some l) such that for all $k \geq m_j$,

$$\sup_{n\in\mathbb{N}} \mathbb{E}\left(\|\mathbf{\Psi}^n\|^2_{X,\sigma_k^M\wedge\tau_n^{M,T}} - \|\mathbf{\Psi}^n\|^2_{X,\sigma_\infty^M\wedge\tau_n^{M,T}}\right) \leq 2^{-2j}.$$

In particular, through a relabeling of $\sigma_{m_j}^M = \sigma_l^M$,

$$\mathbb{E}\left(\|\mathbf{\Psi}^{m_j}\|^2_{X,\sigma_{m_j}^M} - \|\mathbf{\Psi}^{m_j}\|^2_{X,\sigma_\infty^M}\right) \leq 2^{-2j}$$

by choosing n as m_j and using that $\sigma_\infty^M \leq \sigma_k^M \leq \sigma_{m_j}^M \leq \tau_{m_j}^{M,T}$. In a familiar way we define

$$\Omega_j' := \left\{\|\mathbf{\Psi}^{m_j}\|^2_{X,\sigma_{m_j}^M} - \|\mathbf{\Psi}^{m_j}\|^2_{X,\sigma_\infty^M} < 2^{-(j+2)}\right\}$$

so that, just as we showed for (4.20),

$$\check{\Omega}_K := \bigcap_{j=K}^{\infty} \Omega_j', \qquad \check{\Omega} := \bigcup_{K=1}^{\infty} \check{\Omega}_K, \qquad \mathbb{P}\left(\check{\Omega}\right) = 1.$$

For arbitrary given $R > 0$, we find a constant M such that at every $\omega \in \hat{\Omega} \cap \check{\Omega}$, either $\sigma_\infty^M = T$ or $\|\mathbf{\Psi}\|^2_{X,\sigma_\infty^M} \geq R$. In both instances it is clear that $\tau^{R,T} \leq \sigma_\infty^M$, thus proving the proposition. To this end we fix an $\omega \in \hat{\Omega} \cap \check{\Omega}$ such that $\sigma_\infty^M < T$. As σ_∞^M is the decreasing limit of $(\sigma_{m_j}^M)$, then for sufficiently large m_j we must also have that $\sigma_{m_j}^M < T$. Exactly as in (4.22),

$$\left\| \Psi^{m_j} \right\|_{X,\sigma_{m_j}^M} = (\tilde{M} - 1 + 2^{-j}) + \left\| \Psi^{m_j} \right\|_{X,0}. \tag{4.26}$$

From the proven convergence we also have that for sufficiently large m_j,

$$\left\| \Psi - \Psi^{m_j} \right\|_{X,\sigma_\infty^M} < 1, \tag{4.27}$$

which implies that $\| \Psi \|_{X,\sigma_\infty^M} > \| \Psi^{m_j} \|_{X,\sigma_\infty^M} - 1$, and likewise as $\omega \in \check{\Omega}_K$ for some K,

$$\left\| \Psi^{m_j} \right\|_{X,\sigma_{m_j}^M}^2 - \left\| \Psi^{m_j} \right\|_{X,\sigma_\infty^M}^2 < 1. \tag{4.28}$$

We fix an m_j large enough so that (4.26), (4.27), and (4.28) all hold. Substituting (4.26) into (4.28) gives that

$$\left\| \Psi^{m_j} \right\|_{X,\sigma_\infty^M}^2 > \left((\tilde{M} - 1 + 2^{-j}) + \left\| \Psi^{m_j} \right\|_{X,0} \right)^2 - 1 > (\tilde{M} - 1)^2 - 1.$$

If $\tilde{M} > 2$, then the expression on the right is positive and

$$\left\| \Psi^{m_j} \right\|_{X,\sigma_\infty^M} > \left((\tilde{M} - 1)^2 - 1 \right)^{\frac{1}{2}}.$$

Furthermore,

$$\| \Psi \|_{X,\sigma_\infty^M} > \left((\tilde{M} - 1)^2 - 1 \right)^{\frac{1}{2}} - 1,$$

where the right hand side is of course monotone increasing and unbounded in $\tilde{M}$ and hence M. By choosing M large enough such that

$$\left[\left((\tilde{M} - 1)^2 - 1 \right)^{\frac{1}{2}} - 1 \right]^2 > R,$$

we complete the proof. $\qquad\qquad\qquad\qquad\qquad\qquad\qquad\qquad\qquad\qquad\qquad\qquad\square$

4.4   Enhanced Regularity and an Energy Equality

Having established a framework for SPDEs in Chap. 3, we introduce techniques to facilitate our analysis in it. The Itô Formula is well regarded as one of the most useful tools in stochastic analysis, and we establish what can be considered as a specific case of this result in our framework. We shall introduce a new setting in which the established solution framework falls, with the understanding that we would like to apply this to solutions while also using the results to deduce the existence of solutions when they are not a priori known. This is well understood in the typical variational framework, for which we again refer to [51, 57], but we take care in addressing subtle changes. The first is the loss of the duality structure, though for this result we do assume a bilinear form relation which behaves similarly. The second is that we conduct the proof for *local* solutions, necessary for our motivating class of equations, so it is important for us to explicitly address how the localization affects the proof. Indeed, the consideration of local solutions, as well as the related localization in the construction of the integral, martingale theory, and analytical techniques, is an important extension of the framework of [51, 56, 57]. Similarly the final key change is that we do not assume any integrability over the probability space of our processes, demanding again another source of localization which we find worthy of detailing.

To prove this energy equality we shall rely on looking at partitions in time over which some nice properties are satisfied, before taking the limit as the increments go to zero. Toward this, we recall the following lemma from [57, Lemma 4.2.6]. This section follows the ideas of [57, Lemma 4.2.5].

Lemma 4.4 *Let X_1, X_2 be two Banach spaces with continuous embedding $X_1 \hookrightarrow X_2$ and suppose that for some $T > 0$ and a stopping time τ, $\Phi : \Omega \times [0, T] \to X_2$ is such that for $\mathbb{P} - a.e.$ ω, $\Phi_\cdot(\omega) \in C([0, T]; X_2)$ and $\Phi.\mathbb{1}_{\cdot \leq \tau} \in L^2(\Omega \times [0, T]; X_1)$. Then for any $A \subset (0, T)$ with $\lambda(A) = 0$ there exists a sequence of partitions (I_l) such that:*

1. *$I_l := \left\{0 = t_0^l < t_1^l < \cdots < t_{k_l}^l = T\right\}$, $\max_j |t_j^l - t_{j-1}^l| \to 0$ as $l \to \infty$.*
2. *$I_l \subset I_{l+1}$.*
3. *$I_l \cap A = \emptyset$.*
4. *For $\mathbb{P} - a.e.$ ω and every t_j^l with $1 \leq j \leq k_{l-1}$, $\Phi_{t_j^l}(\omega)\mathbb{1}_{t_j^l \leq \tau(\omega)} \in X_1$.*
5. *The processes $\hat{\Phi}^l, \tilde{\Phi}^l$ defined at each $t \in [0, T]$ and $\omega \in \Omega$ by*

$$\hat{\Phi}_t^l(\omega) := \sum_{j=2}^{k_l} \mathbb{1}_{[t_{j-1}^l, t_j^l)}(t)\Phi_{t_{j-1}^l}(\omega)\mathbb{1}_{t_{j-1}^l \leq \tau(\omega)},$$

$$\tilde{\Phi}_t^l(\omega) := \sum_{j=1}^{k_{l-1}} \mathbb{1}_{[t_{j-1}^l, t_j^l)}(t)\Phi_{t_j^l}(\omega)\mathbb{1}_{t_j^l \leq \tau(\omega)}$$

belong to $L^2 (\Omega \times [0, T]; X_1)$ and both converge to $\Phi . \mathbb{1}_{.\leq \tau}$ in this space.

Before moving on, we take a moment to discuss this result. In the statement of [57, Lemma 4.2.6], there is no continuity assumption on Φ, and indeed, this is surplus to requirement; however we want to make explicit that it is genuinely the process Φ taken in item 5 and not some other representative of an equivalence class. The assumptions are of course reminiscent of Definition 3.4 and just as was stressed there that the progressively measurable process in V was not necessarily the continuous process itself but just a $\mathbb{P} \times \lambda - a.s.$ equivalent representation, we are reminded again that elements of $L^2 (\Omega \times [0, T]; X_1)$ are only an equivalence class of $\mathbb{P} \times \lambda - a.s.$ equal functions. Therefore the representations $\hat{\Phi}^l$, $\tilde{\Phi}^l$ are significant as they are piecewise constant in the space X_1.

We impose some additional structure on the established framework to conduct the analysis of this section. We work with a triple of embedded Hilbert spaces

$$V \hookrightarrow H \hookrightarrow U,$$

where the embeddings are continuous, V is assumed dense in H, and there exists a continuous bilinear form $\langle \cdot, \cdot \rangle_{U \times V} : U \times V \to \mathbb{R}$ such that for every $\phi \in H$, $\psi \in V$,

$$\langle \phi, \psi \rangle_{U \times V} = \langle \phi, \psi \rangle_H.$$

We suppose that for some $T > 0$ and stopping time τ:

1. $\Psi_0 \in L^2(\Omega; H)$ is $\mathcal{F}_0$-measurable.
2. $\eta \in L^2 (\Omega \times [0, T]; U))$.
3. $B \in I^H(\mathcal{W})$.
4. $\Psi . \mathbb{1}_{\leq \tau} \in L^2 (\Omega \times [0, T]; V))$ and is progressively measurable in V.
5. The identity

$$\Psi_t = \Psi_0 + \int_0^{t \wedge \tau} \eta_s ds + \int_0^{t \wedge \tau} B_s d\mathcal{W}_s \qquad (4.29)$$

holds $\mathbb{P} - a.s.$ in U for all $t \in [0, T]$.

Remark 4.2 Once more, it is clear from assumption 5 that for $\mathbb{P} - a.e.$ ω, $\Psi(\omega) \in C([0, T]; U)$, but the progressively measurable process assumed in item 4 is only equivalent to this $\Psi . \mathbb{1}_{\leq \tau}$ $\mathbb{P} \times \lambda - a.s..$ It is a slight abuse of notation commonplace in the literature, and we shall follow suit without excessive clarification at each use.

With this structure in place, we first look to deduce some improved regularity on Ψ. We are motivated by applications to the existence theory of solutions, and if η, B were given by functions of Ψ, then note that Ψ is *nearly* a local strong solution of the corresponding SPDE. The missing ingredient is pathwise continuity in H, which we look to deduce from only the given properties.

For this we fix an application of Lemma 4.4 relative to the assumptions laid out above. We take T and τ as in the assumptions, $\mathcal{X}_1 = V$, $\mathcal{X}_2 = U$, $\boldsymbol{\Phi} = \boldsymbol{\Psi}$. From item 4 we have that for $\lambda - a.e.\ t \in [0, T]$, $\mathbb{E}\left(\left\|\boldsymbol{\Psi}_t \mathbb{1}_{t \leq \tau}\right\|_V^2\right) < \infty$, and we take A to be the λ-zero set on which this does not hold. Then $\hat{\boldsymbol{\Psi}}^l$, $\tilde{\boldsymbol{\Psi}}^l$, I^l are defined as in Lemma 4.4, and we define

$$I := \bigcup_{l \in \mathbb{N}} I^l.$$

Lemma 4.5 *We have that*

$$\mathbb{E}\left(\sup_{t \in [0,T]} \|\boldsymbol{\Psi}_t\|_H^2\right) < \infty.$$

Proof For $\mathbb{P} - a.e.\ \omega$ and every $s < t$ with $s, t \in I \cap [0, \tau]/\{T\}$, observe that

$$\left\|\int_s^t B_r d\mathcal{W}_r\right\|_H^2 - \left\|\boldsymbol{\Psi}_t - \boldsymbol{\Psi}_s - \int_s^t B_r d\mathcal{W}_r\right\|_H^2 + 2\left\langle \boldsymbol{\Psi}_s, \int_s^t B_r d\mathcal{W}_r\right\rangle_{\mathcal{H}}$$

$$= \left\|\int_s^t B_r d\mathcal{W}_r\right\|_H^2 - \|\boldsymbol{\Psi}_t - \boldsymbol{\Psi}_s\|_H^2 - \left\|\int_s^t B_r d\mathcal{W}_r\right\|_H^2$$

$$\qquad + 2\left\langle \boldsymbol{\Psi}_t - \boldsymbol{\Psi}_s, \int_s^t B_r d\mathcal{W}_r\right\rangle_{\mathcal{H}} + 2\left\langle \boldsymbol{\Psi}_s, \int_s^t B_r d\mathcal{W}_r\right\rangle_{\mathcal{H}}$$

$$= 2\left\langle \boldsymbol{\Psi}_t, \int_s^t B_r d\mathcal{W}_r\right\rangle_{\mathcal{H}} - \|\boldsymbol{\Psi}_t - \boldsymbol{\Psi}_s\|_H^2$$

$$= 2\left\langle \boldsymbol{\Psi}_t, \boldsymbol{\Psi}_t - \boldsymbol{\Psi}_s - \int_s^t \eta_r dr\right\rangle_{\mathcal{H}} - \|\boldsymbol{\Psi}_t\|_H^2 - \|\boldsymbol{\Psi}_s\|_H^2 + 2\langle \boldsymbol{\Psi}_t, \boldsymbol{\Psi}_s\rangle_H$$

$$= \|\boldsymbol{\Psi}_t\|_H^2 - \|\boldsymbol{\Psi}_s\|_H^2 - 2\left\langle \boldsymbol{\Psi}_t, \int_s^t \eta_r dr\right\rangle_{\mathcal{H}}$$

$$= \|\boldsymbol{\Psi}_t\|_H^2 - \|\boldsymbol{\Psi}_s\|_H^2 - 2\int_s^t \langle \eta_r, \boldsymbol{\Psi}_t\rangle_{U \times V} dr,$$

which we rewrite as the equality

$$\|\boldsymbol{\Psi}_t\|_H^2 - \|\boldsymbol{\Psi}_s\|_H^2 = 2\int_s^t \langle \eta_r, \boldsymbol{\Psi}_t\rangle_{U \times V} dr + 2\left\langle \boldsymbol{\Psi}_s, \int_s^t B_r d\mathcal{W}_r\right\rangle_{\mathcal{H}}$$

$$\qquad + \left\|\int_s^t B_r d\mathcal{W}_r\right\|_H^2 - \left\|\boldsymbol{\Psi}_t - \boldsymbol{\Psi}_s - \int_s^t B_r d\mathcal{W}_r\right\|_H^2.$$

Note that we have had to exclude T from this set, as one does not know if $\mathbf{\Psi}_T \in V$ $\mathbb{P} - a.s.$. Using this equality, for any $l \in \mathbb{N}$ and $t = t_i^l \in I_l \cap (0, \tau]/\{T\}$,

$$\|\mathbf{\Psi}_t\|_H^2 - \|\mathbf{\Psi}_0\|_H^2 = \sum_{j=0}^{i-1} \left(\left\| \mathbf{\Psi}_{t_{j+1}^l} \right\|_H^2 - \left\| \mathbf{\Psi}_{t_j^l} \right\|_H^2 \right)$$

$$= \sum_{j=0}^{i-1} \left(2 \int_{t_j^l}^{t_{j+1}^l} \left\langle \eta_r, \mathbf{\Psi}_{t_{j+1}^l} \right\rangle_{U \times V} dr + 2 \left\langle \mathbf{\Psi}_{t_j^l}, \int_{t_j^l}^{t_{j+1}^l} B_r d\mathcal{W}_r \right\rangle_{\mathcal{H}} \right.$$

$$\left. + \left\| \int_{t_j^l}^{t_{j+1}^l} B_r d\mathcal{W}_r \right\|_H^2 - \left\| \mathbf{\Psi}_{t_{j+1}^l} - \mathbf{\Psi}_{t_j^l} - \int_{t_j^l}^{t_{j+1}^l} B_r d\mathcal{W}_r \right\|_H^2 \right)$$

$$= 2 \sum_{j=0}^{i-1} \left(\int_{t_j^l}^{t_{j+1}^l} \left\langle \eta_r, \mathbf{\Psi}_{t_{j+1}^l} \right\rangle_{U \times V} dr + \int_{t_j^l}^{t_{j+1}^l} \left\langle B_r, \mathbf{\Psi}_{t_j^l} \right\rangle_{\mathcal{H}} d\mathcal{W}_r \right)$$

$$+ \sum_{j=0}^{i-1} \left(\left\| \int_{t_j^l}^{t_{j+1}^l} B_r d\mathcal{W}_r \right\|_H^2 - \left\| \mathbf{\Psi}_{t_{j+1}^l} - \mathbf{\Psi}_{t_j^l} - \int_{t_j^l}^{t_{j+1}^l} B_r d\mathcal{W}_r \right\|_H^2 \right)$$

$$= 2 \sum_{j=0}^{i-1} \left(\int_{t_j^l}^{t_{j+1}^l} \left\langle \eta_r, \tilde{\mathbf{\Psi}}_r^l \right\rangle_{U \times V} dr \right) + 2 \sum_{j=1}^{i-1} \left(\int_{t_j^l}^{t_{j+1}^l} \left\langle B_r, \hat{\mathbf{\Psi}}_r^l \right\rangle_{\mathcal{H}} d\mathcal{W}_r \right)$$

$$+ 2 \int_0^{t_1^l} \left\langle B_r, \mathbf{\Psi}_0 \right\rangle_{\mathcal{H}} d\mathcal{W}_r$$

$$+ \sum_{j=0}^{i-1} \left(\left\| \int_{t_j^l}^{t_{j+1}^l} B_r d\mathcal{W}_r \right\|_H^2 - \left\| \mathbf{\Psi}_{t_{j+1}^l} - \mathbf{\Psi}_{t_j^l} - \int_{t_j^l}^{t_{j+1}^l} B_r d\mathcal{W}_r \right\|_H^2 \right)$$

$$= 2 \int_0^t \left\langle \eta_r, \tilde{\mathbf{\Psi}}_r^l \right\rangle_{U \times V} dr + 2 \int_0^t \left\langle B_r, \hat{\mathbf{\Psi}}_r^l \right\rangle_H d\mathcal{W}_r + 2 \int_0^{t_1^l} \left\langle B_r, \mathbf{\Psi}_0 \right\rangle_H d\mathcal{W}_r$$

$$+ \sum_{j=0}^{i-1} \left(\left\| \int_{t_j^l}^{t_{j+1}^l} B_r d\mathcal{W}_r \right\|_H^2 - \left\| \mathbf{\Psi}_{t_{j+1}^l} - \mathbf{\Psi}_{t_j^l} - \int_{t_j^l}^{t_{j+1}^l} B_r d\mathcal{W}_r \right\|_H^2 \right),$$

$$\tag{4.30}$$

where we have applied Proposition 2.17 and Remark 2.4. In particular we have that

$$\|\mathbf{\Psi}_t\|_H^2 \leq \|\mathbf{\Psi}_0\|_H^2 + 2 \int_0^t \left\langle \eta_r, \tilde{\mathbf{\Psi}}_r^l \right\rangle_{U \times V} dr + 2 \int_0^t \left\langle B_r, \hat{\mathbf{\Psi}}_r^l \right\rangle_H d\mathcal{W}_r$$

$$+ 2 \int_0^{t_1^l} \langle B_r, \boldsymbol{\Psi}_0 \rangle_H d\mathcal{W}_r + \sum_{j=0}^{i-1} \left(\left\| \int_{t_j^l}^{t_{j+1}^l} B_r d\mathcal{W}_r \right\|_H^2 \right).$$

Our goal is to show that

$$\mathbb{E} \left(\max_{t \in I_l \cap (0,\tau]/\{T\}} \| \boldsymbol{\Psi}_t \|_H^2 \right) \le c$$

for some constant c independent of l. To this end, observe that

$$\mathbb{E} \left(\max_{t \in I_l \cap (0,\tau]/\{T\}} \| \boldsymbol{\Psi}_t \|_H^2 \right) \le \mathbb{E} \left(\| \boldsymbol{\Psi}_0 \|_H^2 \right) + 2\mathbb{E} \left(\int_0^{T \wedge \tau} \left| \langle \eta_r, \tilde{\boldsymbol{\Psi}}_r^l \rangle_{U \times V} \right| dr \right)$$

$$+ 2\mathbb{E} \left(\sup_{t \in [0,T]} \left| \int_0^{t \wedge \tau} \langle B_r, \hat{\boldsymbol{\Psi}}_r^l \rangle_H d\mathcal{W}_r \right| \right)$$

$$+ 2\mathbb{E} \left(\sup_{t \in [0,T]} \left| \int_0^{t \wedge \tau} \langle B_r, \boldsymbol{\Psi}_0 \rangle_H d\mathcal{W}_r \right| \right)$$

$$+ \mathbb{E} \left[\max_{t \in I_l \cap (0,\tau]/\{T\}} \sum_{j=0}^{i-1} \left(\left\| \int_{t_j^l}^{t_{j+1}^l} B_r d\mathcal{W}_r \right\|_H^2 \right) \right].$$

We shall treat each term individually. First, we have that

$$2\mathbb{E} \left(\int_0^{T \wedge \tau} \left| \langle \eta_r, \tilde{\boldsymbol{\Psi}}_r^l \rangle_{U \times V} \right| dr \right)$$

$$\le \mathbb{E} \left(\int_0^{T \wedge \tau} \| \eta_r \|_U^2 + \left\| \tilde{\boldsymbol{\Psi}}_r^l \right\|_V^2 dr \right)$$

$$\le \mathbb{E} \left(\int_0^{T \wedge \tau} \| \eta_r \|_U^2 dr \right) + \max_{m \le L} \mathbb{E} \left(\int_0^{T \wedge \tau} \left\| \tilde{\boldsymbol{\Psi}}_r^m \right\|_V^2 dr \right) + 1,$$

where L is taken sufficiently large so that for all $m \ge L$,

$$\mathbb{E} \left(\int_0^{T \wedge \tau} \left| \left\| \tilde{\boldsymbol{\Psi}}_r^m \right\|_V^2 - \| \boldsymbol{\Psi}_r \|_V^2 \right| dr \right) \le \frac{1}{2}.$$

For the first stochastic integral we apply the classical Burkholder–Davis–Gundy Inequality, Theorem A.5, seeing that

$$2\mathbb{E} \left(\sup_{t \in [0,T]} \left| \int_0^{t \wedge \tau} \langle B_r, \hat{\boldsymbol{\Psi}}_r^l \rangle_H d\mathcal{W}_r \right| \right)$$

$$\leq c\mathbb{E}\left(\int_0^{T\wedge\tau}\left\|\left\langle B_r,\hat{\boldsymbol{\Psi}}_r^l\right\rangle_H\right\|^2_{\mathscr{L}^2(\mathfrak{U};\mathbb{R})}dr\right)^{\frac{1}{2}}$$

$$\leq c\mathbb{E}\left(\int_0^{T\wedge\tau}\|B_r\|^2_{\mathscr{L}^2(\mathfrak{U};H)}\left\|\hat{\boldsymbol{\Psi}}_r^l\right\|^2_H dr\right)^{\frac{1}{2}}$$

$$\leq c\mathbb{E}\left(\sup_{r\in[0,T\wedge\tau]}\left\|\hat{\boldsymbol{\Psi}}_r^l\right\|^2_H\int_0^{T\wedge\tau}\|B_r\|^2_{\mathscr{L}^2(\mathfrak{U};H)}dr\right)^{\frac{1}{2}}$$

$$= c\mathbb{E}\left(\sup_{r\in[0,T\wedge\tau]}\left\|\hat{\boldsymbol{\Psi}}_r^l\right\|_H\right)^{\frac{1}{2}}\left(\int_0^{T\wedge\tau}\|B_r\|^2_{\mathscr{L}^2(\mathfrak{U};H)}dr\right)^{\frac{1}{2}}$$

$$= \frac{1}{2}\mathbb{E}\left(\sup_{r\in[0,T\wedge\tau]}\left\|\hat{\boldsymbol{\Psi}}_r^l\right\|^2_H\right)+c\mathbb{E}\left(\int_0^{T\wedge\tau}\|B_r\|^2_{\mathscr{L}^2(\mathfrak{U};H)}dr\right)$$

$$= \frac{1}{2}\mathbb{E}\left(\sup_{t\in I_l\cap(0,\tau]/\{T\}}\|\boldsymbol{\Psi}_t\|^2_H\right)+c\mathbb{E}\left(\int_0^{T\wedge\tau}\|B_r\|^2_{\mathscr{L}^2(\mathfrak{U};H)}dr\right),$$

where c here is a generic constant changing from line to line, independent of l. Putting this together, we now see that

$$\frac{1}{2}\mathbb{E}\left(\max_{t\in I_l\cap(0,\tau]/\{T\}}\|\boldsymbol{\Psi}_t\|^2_H\right)$$

$$\leq \mathbb{E}\left(\|\boldsymbol{\Psi}_0\|^2_H\right)+\mathbb{E}\left(\int_0^{T\wedge\tau}\|\eta_r\|^2_U dr\right)$$

$$+ \max_{m\leq L}\mathbb{E}\left(\int_0^{T\wedge\tau}\left\|\tilde{\boldsymbol{\Psi}}_r^m\right\|^2_V dr\right)+1+c\mathbb{E}\left(\int_0^{T\wedge\tau}\|B_r\|^2_{\mathscr{L}^2(\mathfrak{U};H)}dr\right)$$

$$+ 2\mathbb{E}\left(\sup_{t\in[0,T]}\left|\int_0^t\langle B_r,\boldsymbol{\Psi}_0\rangle_H d\mathcal{W}_r\right|\right)$$

$$+ \mathbb{E}\left[\max_{t\in I_l\cap(0,\tau]/\{T\}}\sum_{j=0}^{i-1}\left(\left\|\int_{t_j^l}^{t_{j+1}^l}B_r d\mathcal{W}_r\right\|^2_H\right)\right].$$

For the stochastic integral involving the initial condition, we can treat this identically to generate the bound

$$2\mathbb{E}\left(\sup_{t\in[0,T]}\left|\int_0^{t\wedge\tau}\langle B_r,\boldsymbol{\Psi}_0\rangle_H d\mathcal{W}_r\right|\right)\leq \mathbb{E}\left(\|\boldsymbol{\Psi}_0\|^2_H\right)+c\mathbb{E}\left(\int_0^{T\wedge\tau}\|B_r\|^2_{\mathscr{L}^2(\mathfrak{U};H)}dr\right).$$

As for the final term, it is clear that the supremum over all partitions can be bounded by taking the partition for $t = t_{k_l}^l = T$. Thus

$$
\mathbb{E}\left[\max_{t \in I_l \cap (0,\tau]/\{T\}} \sum_{j=0}^{i-1} \left(\left\|\int_{t_j^l}^{t_{j+1}^l} B_r d\mathcal{W}_r\right\|_H^2\right)\right] \leq \mathbb{E}\left[\sum_{j=0}^{k_l-1} \left(\left\|\int_{t_j^l}^{t_{j+1}^l} B_r d\mathcal{W}_r\right\|_H^2\right)\right]
$$

$$
= \sum_{j=0}^{k_l-1} \mathbb{E}\left(\left\|\int_{t_j^l}^{t_{j+1}^l} B_r d\mathcal{W}_r\right\|_H^2\right)
$$

$$
= \sum_{j=0}^{k_l-1} \mathbb{E}\left(\int_{t_j^l}^{t_{j+1}^l} \|B_r\|_{\mathscr{L}^2(\mathfrak{U};H)}^2 dr\right)
$$

$$
= \mathbb{E}\left(\int_0^T \|B_r\|_{\mathscr{L}^2(\mathfrak{U};H)}^2 dr\right)
$$

having applied Proposition 2.15. In total then we have that

$$
\mathbb{E}\left(\max_{t \in I_l \cap (0,\tau]/\{T\}} \|\mathbf{\Psi}_t\|_H^2\right) \leq 4\mathbb{E}\left(\|\mathbf{\Psi}_0\|_H^2\right) + 2\mathbb{E}\left(\int_0^{T \wedge \tau} \|\eta_r\|_U^2 dr\right)
$$

$$
+ 2\max_{m \leq L} \mathbb{E}\left(\int_0^{T \wedge \tau} \left\|\tilde{\mathbf{\Psi}}_r^m\right\|_V^2 dr\right) + 2
$$

$$
+ c\mathbb{E}\left(\int_0^T \|B_r\|_{\mathscr{L}^2(\mathfrak{U};H)}^2 dr\right),
$$

which gives a finite bound independent of l. As $I_l \subset I_{l+1}$, then the sequence

$$
\max_{t \in I_l \cap (0,\tau]/\{T\}} \|\mathbf{\Psi}_t\|_H^2
$$

is $\mathbb{P} - a.s.$ monotone increasing in l, so we can apply the Monotone Convergence Theorem to see that

$$
\mathbb{E}\left(\sup_{t \in I \cap (0,\tau]/\{T\}} \|\mathbf{\Psi}_t\|_H^2\right) = \lim_{l \to \infty} \mathbb{E}\left(\max_{t \in I_l \cap (0,\tau]/\{T\}} \|\mathbf{\Psi}_t\|_H^2\right) < \infty. \tag{4.31}
$$

Thus for $\mathbb{P} - a.e.\ \omega$, we have that

$$
\sup_{t \in I \cap (0,\tau(\omega)]/\{T\}} \|\mathbf{\Psi}_t(\omega)\|_H^2 = \tilde{c} < \infty
$$

and $\mathbf{\Psi}_\cdot(\omega) \in C([0,T];U)$. We fix such an ω and any $t \in [0,\tau(\omega)]$. As the mesh of the partitions goes to zero, then there is a sequence of times (t_n) in $I \cap (0,\tau(\omega)]/\{T\}$

such that $t_n \longrightarrow t$. The sequence $\left(\Psi_{t_n}(\omega)\right)$ is uniformly bounded in H so admits a weakly convergent subsequence in this space, to a limit which we call ψ. From the continuous embedding of H into U, this weak convergence also holds in U, but from the continuity of $\Psi.(\omega)$ in U, we have that $\left(\Psi_{t_n}(\omega)\right)$ converges strongly and therefore weakly to $\Psi_t(\omega)$ in U. By the uniqueness of limits in the weak topology, we conclude that $\Psi_t(\omega) = \psi$ and thus belongs to H. Moreover, the weak limit preserves the boundedness in H, so $\|\Psi_t(\omega)\|_H \leq \tilde{c}$. Therefore

$$\sup_{t \in I \cap (0,\tau(\omega)]/\{T\}} \|\Psi_t(\omega)\|_H^2 = \sup_{t \in [0,\tau(\omega)]} \|\Psi_t(\omega)\|_H^2$$

for our fixed ω in a full measure set, and thus $\mathbb{P}-a.s..$ We also know that $\Psi. = \Psi._{\wedge\tau}$ from the identity (4.29), so this equality extends to

$$\sup_{t \in I \cap (0,\tau(\omega)]/\{T\}} \|\Psi_t(\omega)\|_H^2 = \sup_{t \in [0,T]} \|\Psi_t(\omega)\|_H^2.$$

Combining this with (4.31) concludes the proof. $\square$

Having now justified that for $\mathbb{P} - a.e.$ ω and all $t \in [0, T]$ that $\Psi_t(\omega) \in H$, we move on to prove weak continuity in this space.

Lemma 4.6 *For $\mathbb{P} - a.e.$ ω, $\Psi.(\omega)$ is weakly continuous in H.*

Proof We fix $t \in [0, T]$ but now take any sequence of times (t_n) such that $t_n \to t$. For $\mathbb{P} - a.e.$ ω and any given $\phi \in H$ we must justify that

$$\lim_{n \to \infty} \left\langle \Psi_{t_n}(\omega) - \Psi_t(\omega), \phi \right\rangle_H = 0. \tag{4.32}$$

For any $\varepsilon > 0$, from the density of V in H, there exists $\phi^k \in V$ such that

$$\left\| \phi - \phi^k \right\|_H < \frac{\varepsilon}{4\|\Psi(\omega)\|_{L^\infty([0,T];H)}},$$

and then from the continuity in U, there exists an $N \in \mathbb{N}$ sufficiently large such that for all $n \geq N$,

$$\left\| \Psi_{t_n}(\omega) - \Psi_t(\omega) \right\|_U < \frac{\varepsilon}{2\|\phi^k\|_V}.$$

Putting this together,

$$\left| \left\langle \Psi_{t_n}(\omega) - \Psi_t(\omega), \phi \right\rangle_H \right|$$
$$\leq \left| \left\langle \Psi_{t_n}(\omega) - \Psi_t(\omega), \phi - \phi^k \right\rangle_H \right| + \left| \left\langle \Psi_{t_n}(\omega) - \Psi_t(\omega), \phi^k \right\rangle_H \right|$$

$$\leq \left\| \mathbf{\Psi}_{t_n}(\omega) - \mathbf{\Psi}_t(\omega) \right\|_H \left\| \phi - \phi^k \right\|_H + \left| \left\langle \mathbf{\Psi}_{t_n}(\omega) - \mathbf{\Psi}_t(\omega), \phi^k \right\rangle_{U \times V} \right|$$

$$\leq 2 \|\mathbf{\Psi}(\omega)\|_{L^\infty([0,T];H)} \left\| \phi - \phi^k \right\|_H + \left\| \mathbf{\Psi}_{t_n}(\omega) - \mathbf{\Psi}_t(\omega) \right\|_U \left\| \phi^k \right\|_V$$

$$< \varepsilon$$

as required.   □

Guided by the goal to show that $\mathbf{\Psi} \in C([0,T];H)$ $\mathbb{P} - a.s.$, and recalling Lemma 2.2, then continuity of the norm process of $\mathbf{\Psi}$ is now desirable. We show this with an explicit evolution equation, which is the energy equality. We first prove this at times in I:

Lemma 4.7 *The equality*

$$\|\mathbf{\Psi}_t\|_H^2 = \|\mathbf{\Psi}_0\|_H^2 + \int_0^t \left(2\langle \eta_s, \mathbf{\Psi}_s \rangle_{U \times V} + \|B_s\|_{\mathscr{L}^2(\mathcal{U};H)}^2 \right) ds$$

$$+ 2 \int_0^t \langle B_s, \mathbf{\Psi}_s \rangle_H d\mathcal{W}_s \tag{4.33}$$

holds $\mathbb{P} - a.s.$ *in* $\mathbb{R}$ *for all* $t \in I \cap (0, \tau]/\{T\}$.

Proof We recall (4.30), that is,

$$\|\mathbf{\Psi}_t\|_H^2 - \|\mathbf{\Psi}_0\|_H^2 = 2 \int_0^t \left\langle \eta_r, \tilde{\mathbf{\Psi}}_r^l \right\rangle_{U \times V} dr + 2 \int_0^t \left\langle B_r, \hat{\mathbf{\Psi}}_r^l \right\rangle_H d\mathcal{W}_r$$

$$+ 2 \int_0^{t_1^l} \langle B_r, \mathbf{\Psi}_0 \rangle_H d\mathcal{W}_r$$

$$+ \sum_{j=0}^{i-1} \left(\left\| \int_{t_j^l}^{t_{j+1}^l} B_r d\mathcal{W}_r \right\|_H^2 - \left\| \mathbf{\Psi}_{t_{j+1}^l} - \mathbf{\Psi}_{t_j^l} - \int_{t_j^l}^{t_{j+1}^l} B_r d\mathcal{W}_r \right\|_H^2 \right)$$

holds $\mathbb{P} - a.s.$ for all $t = t_i^l \in I_l \cap (0, \tau]/\{T\}$. The idea from now is to fix any $t \in I \cap (0, \tau]/\{T\}$ and consider the limit of this identity (as $l \to \infty$) $\mathbb{P} - a.s.$. In fact limits of the whole sequence will not be possible, but for each term it is sufficient to take a subsequence convergent to what we desire, and to iterate the procedure of taking successive subsequences. For all sufficiently large l the above identity holds at any $t \in I \cap (0, \tau]/\{T\}$, and hence we can consider any such t. We first of all argue that in this limit, the following convergence holds:

$$2 \int_0^t \left\langle \eta_r, \tilde{\mathbf{\Psi}}_r^l \right\rangle_{U \times V} dr \longrightarrow 2 \int_0^t \langle \eta_r, \mathbf{\Psi}_r \rangle_{U \times V} dr \tag{4.34}$$

$$2\int_0^t \left\langle B_r, \hat{\Psi}_r^l \right\rangle_H d\mathcal{W}_r \longrightarrow 2\int_0^t \langle B_r, \Psi_r \rangle_H d\mathcal{W}_r \tag{4.35}$$

$$2\int_0^{t_1^l} \langle B_r, \Psi_0 \rangle_H d\mathcal{W}_r \longrightarrow 0. \tag{4.36}$$

The first term is shown by the convergence in $L^1(\Omega; \mathbb{R})$,

$$\mathbb{E}\left| \int_0^t \left\langle \eta_r, \tilde{\Psi}_r^l - \Psi_r \right\rangle_{U \times V} dr \right| \leq \mathbb{E}\int_0^t \|\eta_r\|_U \left\| \tilde{\Psi}_r^l - \Psi_r \right\|_V dr$$

$$\leq \left(\mathbb{E}\int_0^t \|\eta_r\|_U^2 dr \right)^{\frac{1}{2}} \left(\mathbb{E}\int_0^t \left\| \tilde{\Psi}_r^l - \Psi_r \right\|_V^2 dr \right)^{\frac{1}{2}},$$

which approaches zero from the construction of the sequence $\tilde{\Psi}^l$ (recall Lemma 4.4), thus exhibiting a $\mathbb{P} - a.s.$ convergent subsequence. We verify the limit in the next terms with the Stochastic Dominated Convergence Theorem, Lemma 2.7. In particular, as $\hat{\Psi}^l \to \Psi$ in $L^2(\Omega \times [0, T]; V)$, then there exists a $\mathbb{P} \times \lambda - a.s.$ convergent subsequence in V hence in H. Consequently, observe that

$$\sum_{i=1}^{\infty} \left| \langle B(e_i), \Psi \rangle_H - \left\langle B(e_i), \hat{\Psi}^l \right\rangle_H \right|^2 = \sum_{i=1}^{\infty} \left| \left\langle B(e_i), \Psi - \hat{\Psi}^l \right\rangle_H \right|^2$$

$$\leq \left(\sum_{i=1}^{\infty} \|B(e_i)\|_H^2 \right) \left\| \Psi - \hat{\Psi}^l \right\|_H^2$$

so in fact the corresponding subsequence of $\left(\left\langle B, \hat{\Psi}^l \right\rangle_H \right)$ is $\mathbb{P} \times \lambda - a.s.$ convergent to $\langle B, \Psi \rangle_H$ in $\mathscr{L}^2(\mathfrak{U}; \mathbb{R})$. By the construction of $\hat{\Psi}^l$ from Lemma 4.4, it is evident that

$$\sup_{r \in [0,T]} \left\| \hat{\Psi}_s^l \right\|_H^2 \leq \sup_{r \in [0,T]} \|\Psi_s\|_H^2,$$

which we know to be $\mathbb{P} - a.s.$ finite from Lemma 4.5, so one can choose the dominating function

$$\sup_{r \in [0,T]} \|\Psi_r\|_H^2 \sum_{i=1}^{\infty} \|B(e_i)\|_H^2$$

as Q in Lemma 2.7, facilitating passage to the limit of a subsequence $\mathbb{P} - a.s.$ in (4.35). The same is true of (4.35), by an application of Lemma 2.7 to the subsequence of $\left(\langle B_r, \Psi_0 \rangle_H \mathbb{1}_{r \leq t_1^l} \right)$, which approaches zero by continuity of the integral (Proposition 2.18). We next inspect the limit of

$$\sum_{j=0}^{i-1}\left\|\int_{t_j^l}^{t_{j+1}^l} B_r d\mathcal{W}_r\right\|_H^2,$$

which in Proposition 2.11 was shown to be given, in probability, by

$$\int_0^t \|B_s\|_{\mathscr{L}^2(\mathfrak{U};H)}^2 ds$$

so once more we can extract a subsequence along which $\mathbb{P}-a.s.$ convergence holds. Furthermore, comparing what we have shown to (4.33), it remains to demonstrate that

$$\sum_{j=0}^{i-1}\left\|\Psi_{t_{j+1}^l} - \Psi_{t_j^l} - \int_{t_j^l}^{t_{j+1}^l} B_r d\mathcal{W}_r\right\|_H^2 \longrightarrow 0. \tag{4.37}$$

To this end, we wish to use that

$$\Psi_{t_{j+1}^l} - \Psi_{t_j^l} - \int_{t_j^l}^{t_{j+1}^l} B_r d\mathcal{W}_r = \int_{t_j^l}^{t_{j+1}^l} \eta_r dr \tag{4.38}$$

in U, from the identity (4.29), which appears difficult as the H norm is above the regularity of this integral. We rewrite

$$\left\|\Psi_{t_{j+1}^l} - \Psi_{t_j^l} - \int_{t_j^l}^{t_{j+1}^l} B_r d\mathcal{W}_r\right\|_H^2$$

$$= \left\langle \Psi_{t_{j+1}^l} - \Psi_{t_j^l} - \int_{t_j^l}^{t_{j+1}^l} B_r d\mathcal{W}_r, \Psi_{t_{j+1}^l} - \Psi_{t_j^l} - \int_{t_j^l}^{t_{j+1}^l} B_r d\mathcal{W}_r \right\rangle_H$$

$$= \left\langle \Psi_{t_{j+1}^l} - \Psi_{t_j^l} - \int_{t_j^l}^{t_{j+1}^l} B_r d\mathcal{W}_r, \Psi_{t_{j+1}^l} - \Psi_{t_j^l} \right\rangle_H$$

$$- \left\langle \Psi_{t_{j+1}^l} - \Psi_{t_j^l} - \int_{t_j^l}^{t_{j+1}^l} B_r d\mathcal{W}_r, \int_{t_j^l}^{t_{j+1}^l} B_r d\mathcal{W}_r \right\rangle_H. \tag{4.39}$$

Let us consider the first of these terms. Indeed,

$$\sum_{j=0}^{i-1}\left\langle \Psi_{t_{j+1}^l} - \Psi_{t_j^l} - \int_{t_j^l}^{t_{j+1}^l} B_r d\mathcal{W}_r, \Psi_{t_{j+1}^l} - \Psi_{t_j^l}\right\rangle_H$$

$$=\sum_{j=0}^{i-1}\left\langle \Psi_{t_{j+1}^l} - \Psi_{t_j^l} - \int_{t_j^l}^{t_{j+1}^l} B_r d\mathcal{W}_r, \Psi_{t_{j+1}^l}\right\rangle_H$$

$$-\sum_{j=1}^{i-1}\left\langle \Psi_{t_{j+1}^l} - \Psi_{t_j^l} - \int_{t_j^l}^{t_{j+1}^l} B_r d\mathcal{W}_r, \Psi_{t_j^l}\right\rangle_H -\left\langle \Psi_{t_1^l} - \Psi_0 - \int_0^{t_1^l} B_r d\mathcal{W}_r, \Psi_0\right\rangle_H$$

$$=\sum_{j=0}^{i-1}\left\langle \int_{t_j^l}^{t_{j+1}^l} \eta_r dr, \Psi_{t_{j+1}^l}\right\rangle_{U \times V} -\sum_{j=1}^{i-1}\left\langle \int_{t_j^l}^{t_{j+1}^l} \eta_r dr, \Psi_{t_j^l}\right\rangle_{U \times V}$$

$$-\left\langle \Psi_{t_1^l} - \Psi_0 - \int_0^{t_1^l} B_r d\mathcal{W}_r, \Psi_0\right\rangle_H, \tag{4.40}$$

where we have excluded the case for Ψ_0 as this may not belong to V, in order to apply (4.38) as we have done. Note that

$$\left\langle \int_{t_j^l}^{t_{j+1}^l} \eta_r dr, \Psi_{t_{j+1}^l}\right\rangle_{U \times V} = \int_{t_j^l}^{t_{j+1}^l} \left\langle \eta_r, \Psi_{t_{j+1}^l}\right\rangle_{U \times V} dr$$

$$= \int_0^t \left\langle \eta_r, \Psi_{t_{j+1}^l}\right\rangle_{U \times V} \mathbb{1}_{[t_j^l, t_{j+1}^l]}(r) dr \tag{4.41}$$

and similarly for t_j^l. Moreover, we can rewrite

$$\tilde{\Psi}_r^l \mathbb{1}_{r \leq t} = \sum_{j=0}^{k_l-2} \mathbb{1}_{[t_j^l, t_{j+1}^l]}(r) \Psi_{t_{j+1}^l} \mathbb{1}_{r \leq t}.$$

Therefore, if $i \leq k_{l-1}$, then

$$\sum_{j=0}^{i-1} \mathbb{1}_{[t_j^l, t_{j+1}^l]}(r) \Psi_{t_{j+1}^l} \mathbb{1}_{r \leq t} = \tilde{\Psi}_r^l \mathbb{1}_{r \leq t},$$

but this is certainly the case as $t = t_i^l \in I \cap (0, \tau]/\{T\}$ is excluded from being $T = t_{k_l}^l$. Meanwhile, we have that

$$\hat{\Psi}_r^l \mathbb{1}_{r \leq t} = \sum_{j=1}^{k_l-1} \mathbb{1}_{[t_j^l, t_{j+1}^l]}(r) \Psi_{t_j^l} \mathbb{1}_{r \leq t},$$

which implies

$$\sum_{j=1}^{i-1} \mathbb{1}_{[t_j^l, t_{j+1}^l]}(r) \mathbf{\Psi}_{t_j^l} \mathbb{1}_{r \leq t} = \hat{\mathbf{\Psi}}_r^l \mathbb{1}_{r \leq t}.$$

Together with (4.41), we get that

$$\sum_{j=0}^{i-1} \left\langle \int_{t_j^l}^{t_{j+1}^l} \eta_r dr, \mathbf{\Psi}_{t_{j+1}^l} \right\rangle_{U \times V} = \int_0^t \left\langle \eta_r, \tilde{\mathbf{\Psi}}_r^l \right\rangle_{U \times V} dr$$

$$\sum_{j=1}^{i-1} \left\langle \int_{t_j^l}^{t_{j+1}^l} \eta_r dr, \mathbf{\Psi}_{t_j^l} \right\rangle_{U \times V} = \int_0^t \left\langle \eta_r, \hat{\mathbf{\Psi}}_r^l \right\rangle_{U \times V} dr.$$

Revisiting (4.40), it is apparent that

$$\sum_{j=0}^{i-1} \left\langle \mathbf{\Psi}_{t_{j+1}^l} - \mathbf{\Psi}_{t_j^l} - \int_{t_j^l}^{t_{j+1}^l} B_r d\mathcal{W}_r, \mathbf{\Psi}_{t_{j+1}^l} - \mathbf{\Psi}_{t_j^l} \right\rangle_H$$

$$= \int_0^t \left\langle \eta_r, \tilde{\mathbf{\Psi}}_r^l - \hat{\mathbf{\Psi}}_r^l \right\rangle_{U \times V} dr - \left\langle \mathbf{\Psi}_{t_1^l} - \mathbf{\Psi}_0 - \int_0^{t_1^l} B_r d\mathcal{W}_r, \mathbf{\Psi}_0 \right\rangle_H,$$

which we argue approaches zero along a subsequence $\mathbb{P} - a.s..$ We construct the first term as

$$\int_0^t \left\langle \eta_r, \tilde{\mathbf{\Psi}}_r^l - \hat{\mathbf{\Psi}}_r^l \right\rangle_{U \times V} dr = \int_0^t \left\langle \eta_r, \tilde{\mathbf{\Psi}}_r^l - \mathbf{\Psi}_r \right\rangle_{U \times V} dr - \int_0^t \left\langle \eta_r, \hat{\mathbf{\Psi}}_r^l - \mathbf{\Psi}_r \right\rangle_{U \times V} dr,$$

to be treated exactly as we did for (4.34). The remaining term is split up,

$$\left\langle \mathbf{\Psi}_{t_1^l} - \mathbf{\Psi}_0 - \int_0^{t_1^l} B_r d\mathcal{W}_r, \mathbf{\Psi}_0 \right\rangle_H = \left\langle \mathbf{\Psi}_{t_1^l} - \mathbf{\Psi}_0, \mathbf{\Psi}_0 \right\rangle_H - \left\langle \int_0^{t_1^l} B_r d\mathcal{W}_r, \mathbf{\Psi}_0 \right\rangle_H.$$

The first inner product must approach zero from the weak continuity shown in Proposition 4.6, whereas in the second we may use Proposition 2.17 to take $\mathbf{\Psi}_0$ inside the integral, which was treated in (4.36). So returning to (4.39), with the goal of (4.37) still in mind, we turn our attention to

$$\sum_{j=0}^{i-1} \left\langle \mathbf{\Psi}_{t_{j+1}^l} - \mathbf{\Psi}_{t_j^l} - \int_{t_j^l}^{t_{j+1}^l} B_r d\mathcal{W}_r, \int_{t_j^l}^{t_{j+1}^l} B_r d\mathcal{W}_r \right\rangle_H. \qquad (4.42)$$

For this we cannot use a time approximation in order to utilize (4.38) as we have just done, so we look to consider an approximation in the Hilbert Space instead. From the density of V in H, there exists an orthonormal basis of H consisting of elements from V, which we call (a_k), and let $\mathcal{P}_n$ be the corresponding orthogonal projection in H onto span $\{a_1, \ldots, a_n\}$. Then (4.42) can itself be rewritten as

$$\sum_{j=0}^{i-1} \left\langle \Psi_{t_{j+1}^l} - \Psi_{t_j^l} - \int_{t_j^l}^{t_{j+1}^l} B_r d\mathcal{W}_r, \, \mathcal{P}_n \int_{t_j^l}^{t_{j+1}^l} B_r d\mathcal{W}_r \right\rangle_H \tag{4.43}$$

$$+ \sum_{j=0}^{i-1} \left\langle \Psi_{t_{j+1}^l} - \Psi_{t_j^l} - \int_{t_j^l}^{t_{j+1}^l} B_r d\mathcal{W}_r, \, (I - \mathcal{P}_n) \int_{t_j^l}^{t_{j+1}^l} B_r d\mathcal{W}_r \right\rangle_H \tag{4.44}$$

for any $n \in \mathbb{N}$, where I is the identity operator in H. Of course, the purpose of this is to use (4.38) and rewrite the first term as

$$\sum_{j=1}^{i-1} \left\langle \int_{t_j^l}^{t_{j+1}^l} \eta_r dr, \, \mathcal{P}_n \int_{t_j^l}^{t_{j+1}^l} B_r d\mathcal{W}_r \right\rangle_{U \times V},$$

but we can be more astute here. In fact we appreciate that $\mathcal{P}_n \in \mathscr{L}(H; V)$, so by applying Proposition 2.16 we can take the $\mathcal{P}_n$ through the stochastic integral and see that $\mathcal{P}_n B \in \bar{I}^V(\mathcal{W})$. Thus if we define the V-valued process $M_s := \int_0^s \mathcal{P}_n B_r d\mathcal{W}_r$, then this term is

$$\sum_{j=0}^{i-1} \left\langle \Psi_{t_{j+1}^l} - \Psi_{t_j^l} - \int_{t_j^l}^{t_{j+1}^l} B_r d\mathcal{W}_r, \, M_{t_{j+1}^l} - M_{t_j^l} \right\rangle_H,$$

which can be treated identically to (4.40) (in fact a little more simply, as $M_0 = 0$). Therefore it only remains to consider (4.44), which we control by

$$\left| \sum_{j=0}^{i-1} \left\langle \Psi_{t_{j+1}^l} - \Psi_{t_j^l} - \int_{t_j^l}^{t_{j+1}^l} B_r d\mathcal{W}_r, \, (I - \mathcal{P}_n) \int_{t_j^l}^{t_{j+1}^l} B_r d\mathcal{W}_r \right\rangle_H \right|$$

$$\leq \sum_{j=0}^{i-1} \left\| \Psi_{t_{j+1}^l} - \Psi_{t_j^l} - \int_{t_j^l}^{t_{j+1}^l} B_r d\mathcal{W}_r \right\|_H \left\| (I - \mathcal{P}_n) \int_{t_j^l}^{t_{j+1}^l} B_r d\mathcal{W}_r \right\|_H$$

$$\leq \left(\sum_{j=0}^{i-1} \left\| \Psi_{t_{j+1}^l} - \Psi_{t_j^l} - \int_{t_j^l}^{t_{j+1}^l} B_r d\mathcal{W}_r \right\|_H^2 \right)^{\frac{1}{2}} \left(\sum_{j=0}^{i-1} \left\| (I - \mathcal{P}_n) \int_{t_j^l}^{t_{j+1}^l} B_r d\mathcal{W}_r \right\|_H^2 \right)^{\frac{1}{2}}.$$

Here we have that $I - \mathcal{P}_n \in \mathscr{L}(H; H)$, so this operator can be taken through the integral again by Proposition 2.16, and we once more understand that

$$\sum_{j=0}^{i-1} \left\| (I - \mathcal{P}_n) \int_{t_j^l}^{t_{j+1}^l} B_r dW_r \right\|_H^2 \tag{4.45}$$

, due to Proposition 2.11, converges in probability (and hence $\mathbb{P} - a.s.$ along a subsequence) to

$$\int_0^t \|(I - \mathcal{P}_n)B_s\|_{\mathscr{L}^2(\mathfrak{U};H)}^2 ds.$$

By the Monotone Convergence Theorem, this approaches zero as $n \to \infty$. It is not necessarily clear how to utilize this, so we pause and collect the full picture of everything that we have achieved in the aim of (4.37). We have established that

$$\sum_{j=0}^{i-1} \left\| \Psi_{t_{j+1}^l} - \Psi_{t_j^l} - \int_{t_j^l}^{t_{j+1}^l} B_r dW_r \right\|_H^2$$

$$\leq \left(\sum_{j=0}^{i-1} \left\| \Psi_{t_{j+1}^l} - \Psi_{t_j^l} - \int_{t_j^l}^{t_{j+1}^l} B_r dW_r \right\|_H^2 \right)^{\frac{1}{2}} \left(\sum_{j=0}^{i-1} \left\| \int_{t_j^l}^{t_{j+1}^l} (I - \mathcal{P}_n) B_r dW_r \right\|_H^2 \right)^{\frac{1}{2}}$$

$$+ c_n o_l,$$

where $c_n o_l$ is a random variable which for every fixed n is convergent to zero $\mathbb{P} - a.s.$ along a subsequence as $l \to \infty$. Denoting

$$Y_l := \sum_{j=0}^{i-1} \left\| \Psi_{t_{j+1}^l} - \Psi_{t_j^l} - \int_{t_j^l}^{t_{j+1}^l} B_r dW_r \right\|_H^2,$$

then the above is written as

$$Y_l \leq Y_l^{\frac{1}{2}} \left(\sum_{j=0}^{i-1} \left\| \int_{t_j^l}^{t_{j+1}^l} (I - \mathcal{P}_n) B_r dW_r \right\|_H^2 \right)^{\frac{1}{2}} + c_n o_l.$$

To consider the $\mathbb{P} - a.s.$ limit, again along the iterative taking of subsequences including that required in the limit of (4.45), we first note that we already know this limit of Y_l is well defined; indeed, it is immediate from the equality (4.30) stated at the start of the proof, given that the limit in all other terms has now been established. So, for any fixed n, for the limit $\mathbb{P} - a.s.$ along the subsequence,

$$\lim_{l\to\infty} Y_l \le \lim_{l\to\infty} \left(Y_l^{\frac{1}{2}} \left(\sum_{j=0}^{i-1} \left\| \int_{t_j^l}^{t_{j+1}^l} (I - \mathcal{P}_n) B_r d\mathcal{W}_r \right\|_H^2 \right)^{\frac{1}{2}} \right)$$

$$= \left(\lim_{l\to\infty} Y_l^{\frac{1}{2}} \right) \left(\lim_{l\to\infty} \left(\sum_{j=0}^{i-1} \left\| \int_{t_j^l}^{t_{j+1}^l} (I - \mathcal{P}_n) B_r d\mathcal{W}_r \right\|_H^2 \right)^{\frac{1}{2}} \right)$$

$$= \left(\lim_{l\to\infty} Y_l \right)^{\frac{1}{2}} \int_0^t \|(I - \mathcal{P}_n) B_s\|_{\mathscr{L}^2(\mathfrak{U};H)}^2 ds$$

implying that

$$\left(\lim_{l\to\infty} Y_l \right)^{\frac{1}{2}} \le \int_0^t \|(I - \mathcal{P}_n) B_s\|_{\mathscr{L}^2(\mathfrak{U};H)}^2 ds$$

for any n, which can be made arbitrarily small for sufficiently large n, and hence (4.37) is justified and the proof is concluded. □

It remains to extend Lemma 4.7 to all times in $t \in [0, \tau]$.

Proposition 4.2 *The equality*

$$\|\mathbf{\Psi}_t\|_H^2 = \|\mathbf{\Psi}_0\|_H^2 + \int_0^{t\wedge\tau} \left(2\langle \eta_s, \mathbf{\Psi}_s \rangle_{U\times V} + \|B_s\|_{\mathscr{L}^2(\mathcal{U};H)}^2 \right) ds$$

$$+ 2 \int_0^{t\wedge\tau} \langle B_s, \mathbf{\Psi}_s \rangle_H d\mathcal{W}_s \tag{4.46}$$

holds $\mathbb{P} - a.s.$ *in* $\mathbb{R}$ *for all* $t \in [0, T]$. *Moreover for* $\mathbb{P} - a.e.$ ω, $\mathbf{\Psi}_\cdot(\omega) \in C([0, T]; H)$.

Proof We look to fill in the gaps of the time interval $[0, \tau]$ outside of I in Lemma 4.7 by showing that for any given $t \in [0, \tau]$, and for a sequence of times (t_l) in I convergent to t from below, the corresponding sequence $(\mathbf{\Psi}_{t_l})$ converges to $\mathbf{\Psi}_t$ in H, $\mathbb{P} - a.s..$ Let us fix such a t and construct the required increasing sequence of times (t_l) in I. For each $l \in \mathbb{N}$, there is a unique t_{j-1}^l with the property that $t \in [t_{j-1}^l, t_j^l)$. We consider the sequence of $t_l := t_{j-1}^l$, which is nondecreasing and convergent to t. Let t_m be any element of this sequence. We can consider the process

$$\mathbf{\Phi}_\cdot = \mathbf{\Psi}_\cdot - \mathbf{\Psi}_{t_m},$$

which satisfies the same assumptions as $\mathbf{\Psi}$ and holds the same relation with I. Then by Lemma 4.7, for any t_n with $t_m < t_n$,

$$\left\| \Phi_{t_n} \right\|_H^2 = \left\| \Phi_0 \right\|_H^2 + \int_0^{t_n} \left(2 \langle \eta_s, \Phi_s \rangle_{U \times V} + \left\| B_s \right\|_{\mathscr{L}^2(\mathcal{U};H)}^2 \right) ds + 2 \int_0^{t_n} \langle B_s, \Phi_s \rangle_H d\mathcal{W}_s$$

holds $\mathbb{P} - a.s.$ in $\mathbb{R}$, and subtracting the same equality for t_m,

$$\left\| \Phi_{t_n} \right\|_H^2 = \left\| \Phi_{t_m} \right\|_H^2 + \int_{t_m}^{t_n} \left(2 \langle \eta_s, \Phi_s \rangle_{U \times V} + \left\| B_s \right\|_{\mathscr{L}^2(\mathcal{U};H)}^2 \right) ds + 2 \int_{t_m}^{t_n} \langle B_s, \Phi_s \rangle_H d\mathcal{W}_s.$$

More explicitly, this is

$$\left\| \Psi_{t_n} - \Psi_{t_m} \right\|_H^2 = \int_{t_m}^{t_n} \left(2 \langle \eta_s, \Psi_s - \Psi_{t_m} \rangle_{U \times V} + \left\| B_s \right\|_{\mathscr{L}^2(\mathcal{U};H)}^2 \right) ds$$

$$+ 2 \int_{t_m}^{t_n} \langle B_s, \Psi_s - \Psi_{t_m} \rangle_H d\mathcal{W}_s. \qquad (4.47)$$

If we can show that

$$\lim_{m \to \infty} \sup_{n \geq m} \left\| \Psi_{t_n} - \Psi_{t_m} \right\|_H^2 = 0 \qquad (4.48)$$

$\mathbb{P} - a.s.$, then we would prove that the sequence (Ψ_{t_k}) is $(\mathbb{P} - a.s.)$ Cauchy in H and thus exhibits a limit in H. This limit also holds in U, a space in which Ψ is continuous from the assumed identity (4.29), and thus we know that the limit exists in U and is given by Ψ_t. This must agree with the limit in the finer topology of H, so the convergence of (Ψ_{t_k}) to Ψ_t in H would be verified through the property (4.48). We recall that $\hat{\Psi}^m$ is constant on every interval of the form $[t_{j-1}^m, t_j^m)$, given by $\Psi_{t_{j-1}^m}$, and by construction of the times (t_l) then for some j, $t_m = t_{j-1}^m$ and $t_n \in [t_{j-1}^m, t_j^m)$. In particular, $\hat{\Psi}^m$ is constant on $[t_m, t_n)$ and is given by Ψ_{t_m}. Therefore we can rewrite (4.47) as

$$\left\| \Psi_{t_n} - \Psi_{t_m} \right\|_H^2 = \int_{t_m}^{t_n} \left(2 \langle \eta_s, \Psi_s - \hat{\Psi}_s^m \rangle_{U \times V} + \left\| B_s \right\|_{\mathscr{L}^2(\mathcal{U};H)}^2 \right) ds$$

$$+ 2 \int_{t_m}^{t_n} \langle B_s, \Psi_s - \hat{\Psi}_s^m \rangle_H d\mathcal{W}_s.$$

The required convergence of each of the terms on the right hand side for (4.48) is now conducted exactly as we did for (4.34), (4.35) and is simple by continuity of the integral for the $\int_{t_m}^{t_n} \left\| B_s \right\|_{\mathscr{L}^2(\mathcal{U};H)}^2 ds$ term. Moreover, for a subsequence at least, we have that (Ψ_{t_l}) converges to (Ψ_t) in H $\mathbb{P} - a.s.$, and hence the norms in H converge as well. Continuity of the integrals in the corresponding equalities (4.33) establishes that

$$\|\mathbf{\Psi}_t\|_H^2 = \|\mathbf{\Psi}_0\|_H^2 + \int_0^t \left(2\langle \eta_s, \mathbf{\Psi}_s \rangle_{U \times V} + \|B_s\|_{\mathscr{L}^2(\mathscr{U};H)}^2 \right) ds + 2\int_0^t \langle B_s, \mathbf{\Psi}_s \rangle_H d\mathcal{W}_s,$$

where we recall t was arbitrarily chosen in $[0, \tau]$. Passage to (4.46) is now trivial as $\mathbf{\Psi}_t = \mathbf{\Psi}_\tau$ for $t \geq \tau$. The final part of the result which needs to be proven is the continuity of $\mathbf{\Psi}$ in H. From the equality (4.46) we know that the $\|\mathbf{\Psi}\|_H^2$ is $\mathbb{P} - a.s.$ continuous in $\mathbb{R}$, and from Lemma 4.6 that $\mathbf{\Psi}$ is $\mathbb{P} - a.s.$ weakly continuous in H. The required continuity thus follows from Lemma 2.2. $\square$

We wish to relax the integrability constraints over Ω, in accordance with Definition 3.4. In the same setting of the Hilbert Spaces V, H, U, we impose the new assumptions for some $T > 0$ and stopping time τ:

1. $\mathbf{\Psi}_0 : \Omega \to H$ is $\mathcal{F}_0$-measurable.
2. For $\mathbb{P} - a.e.\ \omega$, $\eta(\omega) \in L^2([0, T]; U)$.
3. $B \in \bar{I}^H(\mathcal{W})$.
4. For $\mathbb{P} - a.e.\ \omega$, $\mathbf{\Psi}.(\omega)\mathbb{1}_{.\leq\tau(\omega)} \in L^2([0, T]; V)$ and $\mathbf{\Psi}.\mathbb{1}_{.\leq\tau}$ is progressively measurable in V.
5. The identity

$$\mathbf{\Psi}_t = \mathbf{\Psi}_0 + \int_0^{t\wedge\tau} \eta_s ds + \int_0^{t\wedge\tau} B_s d\mathcal{W}_s \tag{4.49}$$

holds $\mathbb{P} - a.s.$ in U for all $t \in [0, T]$.

We restate Proposition 4.2 for the new setting.

Proposition 4.3 *The equality*

$$\|\mathbf{\Psi}_t\|_H^2 = \|\mathbf{\Psi}_0\|_H^2 + \int_0^{t\wedge\tau} \left(2\langle \eta_s, \mathbf{\Psi}_s \rangle_{U\times V} + \|B_s\|_{\mathscr{L}^2(\mathscr{U};H)}^2 \right) ds$$
$$+ 2\int_0^{t\wedge\tau} \langle B_s, \mathbf{\Psi}_s \rangle_H d\mathcal{W}_s \tag{4.50}$$

holds $\mathbb{P} - a.s.$ *in* $\mathbb{R}$ *for any* $t \in [0, T]$. *Moreover for* $\mathbb{P} - a.e.\ \omega$, $\mathbf{\Psi}.(\omega) \in C([0, T]; H)$.

Proof The idea is simply to apply Proposition 4.2 for some truncated versions of the processes. We consider the stopping times

$$\tau_n^1 := n \wedge \inf\{0 \leq t < \infty : \int_0^t \|\eta_s\|_U^2 ds \geq n\}$$

$$\tau_n^2 := n \wedge \inf\{0 \leq t < \infty : \int_0^t \|B_s\|_{\mathscr{L}^2(\mathfrak{U};H)}^2 ds \geq n\}$$

$$\tau_n^3 := n \wedge \inf\{0 \le t < \infty : \int_0^t \left\| \mathbf{\Psi}_s \mathbb{1}_{s \le \tau} \right\|_V^2 ds \ge n\}$$

$$\tau_n := \tau_n^1 \wedge \tau_n^2 \wedge \tau_n^3.$$

Then for every n, we have that

$$\mathbf{\Psi}_0 \mathbb{1}_{\|\mathbf{\Psi}_0\|_H \le n}, \qquad \eta_{\cdot} \mathbb{1}_{\cdot \le \tau_n} \mathbb{1}_{\|\mathbf{\Psi}_0\|_H \le n}$$

$$B_{\cdot} \mathbb{1}_{\cdot \le \tau_n} \mathbb{1}_{\|\mathbf{\Psi}_0\|_H \le n}, \qquad \mathbf{\Psi}_{\cdot \wedge \tau_n} \mathbb{1}_{\|\mathbf{\Psi}_0\|_H \le n}$$

satisfy the previous assumptions of 1, 2, 3, 4. Moreover from (4.49) we have that for any $t \in [0, T]$

$$\mathbf{\Psi}_{t \wedge \tau_n} = \mathbf{\Psi}_0 + \int_0^{t \wedge \tau \wedge \tau_n} \eta_s ds + \int_0^{t \wedge \tau \wedge \tau_n} B_s d\mathcal{W}_s$$

$$= \mathbf{\Psi}_0 + \int_0^{t \wedge \tau} \eta_s \mathbb{1}_{s \le \tau_n} ds + \int_0^{t \wedge \tau} B_s \mathbb{1}_{s \le \tau_n} d\mathcal{W}_s.$$

$\mathbb{P} - a.s.$, and moreover

$$\mathbf{\Psi}_{t \wedge \tau_n} \mathbb{1}_{\|\mathbf{\Psi}_0\|_H \le n} = \mathbf{\Psi}_0 \mathbb{1}_{\|\mathbf{\Psi}_0\|_H \le n} + \mathbb{1}_{\|\mathbf{\Psi}_0\|_H \le n} \int_0^{t \wedge \tau} \eta_s \mathbb{1}_{s \le \tau_n} ds$$

$$+ \mathbb{1}_{\|\mathbf{\Psi}_0\|_H \le n} \int_0^{t \wedge \tau} B_s \mathbb{1}_{s \le \tau_n} d\mathcal{W}_s$$

$$= \mathbf{\Psi}_0 \mathbb{1}_{\|\mathbf{\Psi}_0\|_H \le n} + \int_0^{t \wedge \tau} \eta_s \mathbb{1}_{s \le \tau_n} \mathbb{1}_{\|\mathbf{\Psi}_0\|_H \le n} ds$$

$$+ \int_0^{t \wedge \tau} B_s \mathbb{1}_{s \le \tau_n} \mathbb{1}_{\|\mathbf{\Psi}_0\|_H \le n} d\mathcal{W}_s$$

having applied Proposition 2.17. Therefore we can apply Proposition 4.2 to see that the equality

$$\left\| \mathbf{\Psi}_{t \wedge \tau_n} \mathbb{1}_{\|\mathbf{\Psi}_0\|_H \le n} \right\|_H^2 = \left\| \mathbf{\Psi}_0 \mathbb{1}_{\|\mathbf{\Psi}_0\|_H \le n} \right\|_H^2$$

$$+ \int_0^{t \wedge \tau} 2 \langle \eta_s \mathbb{1}_{s \le \tau_n} \mathbb{1}_{\|\mathbf{\Psi}_0\|_H \le n}, \mathbf{\Psi}_{s \wedge \tau_n} \mathbb{1}_{\|\mathbf{\Psi}_0\|_H \le n} \rangle_{U \times V} ds$$

$$+ \int_0^{t \wedge \tau} \left\| B_s \mathbb{1}_{s \le \tau_n} \mathbb{1}_{\|\mathbf{\Psi}_0\|_H \le n} \right\|_{\mathscr{L}^2(\mathcal{U}; H)}^2 ds$$

$$+ 2 \int_0^{t \wedge \tau} \langle B_s \mathbb{1}_{s \le \tau_n} \mathbb{1}_{\|\mathbf{\Psi}_0\|_H \le n}, \mathbf{\Psi}_{s \wedge \tau_n} \mathbb{1}_{\|\mathbf{\Psi}_0\|_H} \rangle_H d\mathcal{W}_s$$

holds for all $t \in [0, T]$ $\mathbb{P} - a.s..$ We rewrite this as

$$\mathbb{1}_{\|\Psi_0\|_H \le n} \left\| \Psi_{t \wedge \tau_n} \right\|_H^2 = \mathbb{1}_{\|\Psi_0\|_H \le n} \|\Psi_0\|_H^2 + \mathbb{1}_{\|\Psi_0\|_H \le n} \int_0^{t \wedge \tau \wedge \tau_n} 2\langle \eta_s, \Psi_s \rangle_{U \times V} ds$$

$$+ \mathbb{1}_{\|\Psi_0\|_H \le n} \int_0^{t \wedge \tau \wedge \tau_n} \| B_s \|_{\mathscr{L}^2(\mathcal{U}; H)}^2 ds$$

$$+ \mathbb{1}_{\|\Psi_0\|_H \le n} 2 \int_0^{t \wedge \tau \wedge \tau_n} \langle B_s, \Psi_s \rangle_H dW_s.$$

Therefore for any $t \in [0, T]$, $\mathbb{P} - a.e.$ ω with n sufficiently large so that $\|\Psi_0(\omega)\|_H \le n$ and $t \le \tau_n(\omega)$, the identity (4.49) holds. We can always find such a large enough n, which completes the justification of this identity. The continuity then follows identically as we have again from Proposition 4.2 that for every n and $\mathbb{P} - a.e.$ ω, $\Psi_{\cdot \wedge \tau_n}(\omega) \in C([0, T]; H)$, and we conclude the proof. □

4.5 SPDEs with Constant Multiplicative Noise

Many techniques in proving existence and uniqueness of an SPDE in this framework rely on simplifying the equation to one where we can apply the standard theory and then constructing solutions in the original framework via some appropriate limit of solutions to the simplified equations. As such, we shall briefly consider a special type of equation in this framework which reduces the driving noise from something infinite dimensional to one dimensional. For this we work again with an arbitrary Hilbert space $\mathcal{H}$.

Proposition 4.4 *Suppose that* $\Psi \in \bar{I}^{\mathcal{H}}$ *and that the operator* $G = (G_i) : \mathfrak{U} \times \mathcal{H} \to \mathcal{H}$ *is such that*

$$G_i : \phi \mapsto \lambda_i \phi$$

for each i *with* $\lambda_i \in \mathbb{R}$, *and*

$$\sum_{i=1}^\infty \lambda_i^2 < \infty. \tag{4.51}$$

Then $G\Psi \in \bar{I}^{\mathcal{H}}(W)$, *and there exists a real valued Brownian Motion* W *such that*

$$\int_0^t G\Psi_s dW_s = \left(\sum_{i=1}^\infty \lambda_i^2 \right)^{1/2} \int_0^t \Psi_s dW_s \tag{4.52}$$

$\mathbb{P} - a.s.$ *for all* $t \ge 0$.

In order to prove the above, we use an intermediary lemma.

Lemma 4.8 *In the setting of Proposition 4.4 the infinite sum*

$$M_s := \sum_{i=1}^{\infty} \lambda_i W_s^i \tag{4.53}$$

is convergent in $L^2(\Omega; \mathbb{R})$ at every s, and the limiting martingale has the representation

$$M = \left(\sum_{i=1}^{\infty} \lambda_i^2 \right)^{1/2} W \tag{4.54}$$

for some real valued Brownian Motion W, $\mathbb{P} - a.s.$ for all $t \geq 0$.

Proof of Lemma 4.8 First, let us verify that the convergence in (4.53) does indeed hold, which is immediate from observing that

$$\sum_{i=1}^{\infty} \lambda_i W_s^i = \sum_{i=1}^{\infty} \int_0^s \lambda_i dW_r^i,$$

which is simply the stochastic integral

$$\int_0^s P(r) d\mathcal{W}_r$$

for the process $P \in \mathcal{I}^{\mathbb{R}}(\mathcal{W})$ defined by $P_{e_i}(r) = \lambda_i$. So M is a continuous genuine martingale, which we show is of the form (4.54) through Levy's Characterization of Brownian Motion, Theorem A.4. Indeed, the quadratic variation process $[M]$ is deduced from Lemma 2.5, where our approximating sequence of martingales

$$M^n = \sum_{i=1}^{n} \lambda_i W_s^i$$

has quadratic variation

$$[M^n]_s = \sum_{i=1}^{n} \lambda_i^2 s$$

which of course converges in $L^1(\Omega; \mathbb{R})$ to the infinite sum, from which we conclude

$$[M]_s = \sum_{i=1}^{\infty} \lambda_i^2 s.$$

Therefore

$$\left[\frac{M}{\left(\sum_{i=1}^{\infty} \lambda_i^2 \right)^{1/2}} \right]_s = s,$$

and we immediately deduce the representation (4.54) from Levy's characterization.

□

Proof of Proposition 4.4 As $\mathcal{G} : \mathcal{H} \to \mathcal{L}^2(\mathcal{U}; \mathcal{H})$ is linear and also bounded from the observation that

$$\| \mathcal{G} \phi \|_{\mathcal{L}^2(\mathcal{U}; \mathcal{H})}^2 = \sum_{i=1}^{\infty} \| \lambda_i \phi \|_{\mathcal{H}}^2 = \sum_{i=1}^{\infty} \lambda_i^2 \| \phi \|_{\mathcal{H}}^2,$$

then it is continuous as a mapping between these spaces, so it preserves the progressive measurability, and evidently the required boundedness to deduce that $\mathcal{G} \Psi \in \bar{I}^{\mathcal{H}}(\mathcal{W})$. To show the identity (4.52), let us rewrite

$$\int_0^t \mathcal{G} \Psi_s d\mathcal{W}_s = \lim_{n \to \infty} \sum_{i=1}^n \int_0^t \lambda_i \Psi_s dW_s^i = \lim_{n \to \infty} \int_0^t \Psi_s dM_s^n$$

with notation M^n as in Proposition 4.8, understanding once more that this limit is taken in $L^2(\Omega; \mathcal{H})$ for the stopped integrals. The localizing stopping times (τ_m) defined by

$$\tau_m = m \wedge \inf\{ 0 \le t < \infty : \int_0^t \sum_{i=1}^{\infty} \lambda_i^2 \| \Psi_s \|_{\mathcal{H}}^2 ds \ge m \}$$

are precisely as in Definition 2.22 and (2.10) to define the left and right sides of (4.52), respectively. It is therefore sufficient to show that for any m,

$$\lim_{n \to \infty} \int_0^t \Psi_s \mathbb{1}_{s \le \tau_m} dM_s^n = \int_0^t \Psi_s \mathbb{1}_{s \le \tau_m} dM_s$$

having simply inserted the representation (4.54) into our required identity. In other words we want that

$$\mathbb{E} \left\| \int_0^t \Psi_s \mathbb{1}_{s \le \tau_m} dM_s - \int_0^t \Psi_s \mathbb{1}_{s \le \tau_m} dM_s^n \right\|_{\mathcal{H}}^2 \longrightarrow 0$$

as $n \to \infty$, which is equivalent to the statement

$$\mathbb{E}\left\| \int_0^t \boldsymbol{\Psi}_s \mathbb{1}_{s \leq \tau_m} d(M - M^n)_s \right\|_{\mathcal{H}}^2 \longrightarrow 0.$$

The same arguments of Proposition 4.8 afford us that the martingale $M - M^n$ which is given at each time s by

$$(M - M_n)_s = \sum_{i=n+1}^{\infty} \lambda_i W_s^i$$

has the representation

$$M - M^n = \left(\sum_{i=n+1}^{\infty} \lambda_i^2 \right)^{1/2} V$$

for a standard Brownian Motion V. So we have that

$$\mathbb{E}\left\| \int_0^t \boldsymbol{\Psi}_s \mathbb{1}_{s \leq \tau_m} d(M - M^n)_s \right\|_{\mathcal{H}}^2 = \mathbb{E}\left\| \left(\sum_{i=n+1}^{\infty} \lambda_i^2 \right)^{1/2} \int_0^t \boldsymbol{\Psi}_s \mathbb{1}_{s \leq \tau_m} dV_s \right\|_{\mathcal{H}}^2$$

$$= \left(\sum_{i=n+1}^{\infty} \lambda_i^2 \right) \mathbb{E}\left\| \int_0^t \boldsymbol{\Psi}_s \mathbb{1}_{s \leq \tau_m} dV_s \right\|_{\mathcal{H}}^2$$

$$= \left(\sum_{i=n+1}^{\infty} \lambda_i^2 \right) \mathbb{E} \int_0^t \left\| \boldsymbol{\Psi}_s \mathbb{1}_{s \leq \tau_m} \right\|_{\mathcal{H}}^2 ds$$

having used the Itô Isometry Corollary 2.1. By definition of the stopping time, the integral is bounded uniformly in ω ($\mathbb{P}-a.s.$) and hence has finite expectation, so we conclude that this approaches zero in the limit from the fact that $\sum_{i=1}^{\infty} \lambda_i^2 < \infty$. $\square$

Appendix A

Here we present miscellaneous results, from both classical and recent literature, which are used throughout the text.

A.1 Classical Results from the Real Valued Theory

We collect some results from the finite dimensional theory, which we employ in our constructions.

Proposition A.1 *Let (M^n) be a sequence in $\mathcal{M}_c^2$ and M a process with values in $\mathbb{R}$ such that at every time $t \geq 0$,*

$$\lim_{n \to \infty} \mathbb{E}\left(|M_t^n - M_t|^2\right) = 0.$$

Then $M \in \mathcal{M}_c^2$.

Proof See [44, Proposition 1.5.23]. □

Proposition A.2 (First Submartingale Inequality) *Let M be a real valued continuous submartingale such that for every $t \geq 0$, $M_t \geq 0 \, \mathbb{P} - a.s..$ Then*

$$\mathbb{P}\left(\left\{\omega \in \Omega : \sup_{t \in [0,T]} M_t(\omega) > \varepsilon \right\}\right) \leq \frac{1}{\varepsilon} \mathbb{E}\left(M_T\right).$$

Proof See [44, Theorem 1.3.8]. □

Theorem A.1 (Doob–Meyer Decomposition) *Let $M \in \mathcal{M}_c^2$. Then there exists a unique (up to indistinguishability) continuous, adapted, nondecreasing real valued process $[M]$ with $[M]_0 = 0 \, (\mathbb{P} - a.s.)$ such that*

© The Author(s), under exclusive license to Springer Nature Switzerland AG 2024
D. Goodair, D. Crisan, *Stochastic Calculus in Infinite Dimensions and SPDEs*,
SpringerBriefs in Mathematics, https://doi.org/10.1007/978-3-031-69586-5

$$M^2 - [M]$$

is a real valued martingale.

Proof See [44, Theorem 1.4.10].   □

Theorem A.2 *Let* $M, N \in \mathcal{M}_c$. *Then there exists a unique (up to indistinguishability) continuous, adapted, bounded-variation process* $[M, N]$ *with* $[M, N]_0 = 0$ *(*$\mathbb{P} - a.s.$*) such that*

$$MN - [M, N]$$

is a real valued martingale.

Proof See [44, Theorem 1.5.13].   □

Lemma A.1 *Let* $M \in \mathcal{M}_c^2$ *be of bounded-variation. Then* M *is constant* $\mathbb{P} - a.s.$

Proof This constitutes part of the proof of [44, Theorem 1.5.13].   □

Lemma A.2 *Let* $M, N \in \mathcal{M}_c^2$. *Then* $\mathbb{P} - a.s.$ *for any* $t \geq 0$,

$$[M, N]_t^2 \leq [M]_t [N]_t.$$

Proof See [44, Problem 1.5.7].   □

Lemma A.3 *Let* $M, N \in \mathcal{M}_c^2$. *Then* $\mathbb{P} - a.s.$ *for any* $T \geq 0$,

$$V_{\mathbb{R}}^T ([M, N]) \leq \frac{1}{2} ([M]_t + [N]_t).$$

Proof See [44, Problem 1.5.7].   □

Theorem A.3 *Let* $M \in \mathcal{M}_c^2$, *and consider any sequence of partitions*

$$I_l := \left\{ 0 = t_0^l < t_1^l < \cdots < t_{k_l}^l = T \right\}$$

with $\max_j |t_j^l - t_{j-1}^l| \to 0$ *as* $l \to \infty$. *Then for all* $t \in [0, T]$, *for any* $\varepsilon > 0$,

$$\lim_{l \to \infty} \mathbb{P} \left(\left\{ \left| \sum_{t_{j+1}^l \leq t} \left| M_{t_{j+1}^l} - M_{t_j^l} \right|^2 - [M]_t \right| > \varepsilon \right\} \right) = 0.$$

If, in addition, $|M|, [M] \in L^\infty (\Omega \times [0, T]; \mathbb{R})$, *then*

$$\lim_{l \to \infty} \mathbb{E} \left(\left| \sum_{t^l_{j+1} \leq t} \left| M_{t^l_{j+1}} - M_{t^l_j} \right|^2 - [M]_t \right| \right) = 0.$$

Proof See [44, Theorem 1.5.8]. □

Theorem A.4 (Lévy's Characterization of Brownian Motion) *Let M be a real valued continuous local martingale with $M_0 = 0 \, \mathbb{P} - a.s.$, and $[M]_t = t$. Then M is a Brownian Motion.*

Proof See [44, Theorem 1.3.16]. □

Theorem A.5 (Burkholder–Davis–Gundy Inequality) *For every $p \geq 1$, there exists a constant C_p such that, for every real valued continuous local martingale M with $M_0 = 0 \, \mathbb{P} - a.s.$, and for any stopping time $\tau \geq 0$,*

$$\mathbb{E} \left(\sup_{t \in [0, \tau]} |M_t|^p \right) \leq C_p \mathbb{E} \left([M]_\tau^{\frac{p}{2}} \right).$$

Proof See [10]. □

A.2   Classical Tightness Criteria

Theorem A.6 *Let (Ψ^n) be a sequence of processes in $\mathcal{D}([0, T]; \mathbb{R})$. Suppose that for any sequence of stopping times (γ_n), $\gamma_n : \Omega \to [0, T]$, and constants (δ_n), $\delta_n \geq 0$, and $\delta_n \to 0$ as $n \to \infty$:*

1. *For every $t \in [0, T]$, the sequence of the laws of (Ψ^n_t) is tight in the space of probability measures over $\mathbb{R}$.*
2. *For every $\varepsilon > 0$, $\lim_{n \to \infty} \mathbb{P} \left(\left\{ \omega \in \Omega : \left| \Psi^n_{(\gamma_n + \delta_n) \wedge T} - \Psi^n_{\gamma_n} \right| > \varepsilon \right\} \right) = 0$.*

Then the sequence of the laws of (Ψ^n) is tight in the space of probability measures over $\mathcal{D}([0, T]; \mathbb{R})$.

Proof See [3, Theorem 1]. □

Theorem A.7 *Let E be a metric space and $\mathbb{F}$ a collection of functions in $C(E; \mathbb{R})$ with the property that $\mathbb{F}$ separates points in E and is closed under addition. Let (μ^n) be a sequence of probability measures on $\mathcal{D}([0, T]; E)$ satisfying the following:*

1. *For each $\varepsilon > 0$, there exists a compact $K \subset E$ such that for every $n \in \mathbb{N}$, $\mu^n (\mathcal{D}([0, T]; K)) > 1 - \varepsilon$.*
2. *For every $f \in \mathbb{F}$, defining the mapping $\tilde{f} : \mathcal{D}([0, T]; E) \to \mathcal{D}([0, T]; \mathbb{R})$ by $[\tilde{f}(\phi)](t) = f[\phi(t)]$, then the sequence $(\mu^n \circ \tilde{f}^{-1})$ is tight in the space of probability measures over $\mathcal{D}([0, T]; \mathbb{R})$.*

Then the sequence (μ^n) is tight in the space of probability measures over $\mathcal{D}([0, T]; E)$.

Proof See [43, Theorem 3.1].   □

A.3   Stochastic Grönwall Lemma

Continuing to look at techniques from PDE theory, a Stochastic Grönwall Lemma will prove of great significance in applications. While in some situations we can apply the classical Grönwall Lemma to the expectation of the process, this is complicated when we have control by the expectation of a product of processes. To overcome this Glatt-Holtz and Ziane proved the following:

Lemma A.4 *Fix $t > 0$ and suppose that ϕ, ψ, η are real valued, nonnegative stochastic processes. Assume, moreover, that there exist constants $c', \hat{c}$ (allowed to depend on t) such that for $\mathbb{P} - a.e.\ \omega$,*

$$\int_0^t \eta_s(\omega)ds \le c', \tag{A.1}$$

and for all stopping times $0 \le \theta_j < \theta_k \le t$,

$$\mathbb{E}\left(\sup_{r\in[\theta_j,\theta_k]} \phi_r\right) + \mathbb{E}\left(\int_{\theta_j}^{\theta_k} \psi_s ds\right) \le \hat{c}\mathbb{E}\left(\left(\phi_{\theta_j} + 1\right) + \int_{\theta_j}^{\theta_k} \eta_s\phi_s ds\right) < \infty.$$

Then there exists a constant C dependent only on $c', \hat{c}, t$ such that

$$\mathbb{E}\left(\sup_{r\in[0,t]} \phi_r\right) + \mathbb{E}\left(\int_0^t \psi_s ds\right) \le C\left[\mathbb{E}(\phi_0) + 1\right].$$

Proof See [32, Lemma 5.3].   □

References

1. Agresti, A.: Delayed blow-up and enhanced diffusion by transport noise for systems of reaction–diffusion equations. Stochastics Partial Differ. Equ. Anal. Comput., 1–75 (2023)
2. Agresti, A.: Global smooth solutions by transport noise of 3D Navier-Stokes equations with small hyperviscosity. arXiv preprint arXiv:2406.09267 (2024)
3. Aldous, D.: Stopping times and tightness. Ann. Probab., 335–340 (1978)
4. Alonso-Orán, D., Bethencourt de León, A.: On the well-posedness of stochastic Boussinesq equations with transport noise. J. Nonlinear Sci. **30**(1), 175–224 (2020)
5. Bagnara, M.: A suitable nonlinear Stratonovich noise prevents blow-up in the Euler equations and other SPDEs. arXiv preprint arXiv:2312.10446 (2023)
6. Bagnara, M., Galeati, L., Maurelli, M.: Regularization by rough Kraichnan noise for the generalised SQG equations. arXiv preprint arXiv:2405.12181 (2024)
7. Billingsley, P.: Convergence of Probability Measures. Wiley (2013)
8. Black, F., Scholes, M.: The pricing of options and corporate liabilities. J. Polit. Econ. **81**(3), 637–654 (1973)
9. Brzeźniak, Z., Capiński, M., Flandoli, F.: Stochastic Navier-Stokes equations with multiplicative noise. Stoch. Anal. Appl. **10**(5), 523–532 (1992)
10. Burkholder, D.L., Davis, B.J., Gundy, R.F.: Integral inequalities for convex functions of operators on martingales. In: Proceedings of the Sixth Berkeley Symposium on Mathematical Statistics and Probability, vol. 2, pp. 223–240. University of California Press Berkeley, California (1972)
11. Butori, F., Flandoli, F., Luongo, E.: On the Itô-Stratonovich diffusion limit for the magnetic field in a 3D thin domain. arXiv preprint arXiv:2401.15701 (2024)
12. Carigi, G., Luongo, E.: Dissipation properties of transport noise in the two-layer quasi-geostrophic model. J. Math. Fluid Mech. **25**(2), 28 (2023)
13. Coffey, W., Kalmykov, Y.P.: The Langevin Equation: With Applications to Stochastic Problems in Physics, Chemistry and electrical Engineering, vol. 27. World Scientific (2012)
14. Conway, J.B.: A Course in Operator Theory. American Mathematical Soc. (2000)
15. Crisan, D., Flandoli, F., Holm, D.D.: Solution properties of a 3D stochastic Euler fluid equation. J. Nonlinear Sci. **29**(3), 813–870 (2019)
16. Crisan, D., Holm, D., Korn, P.: An implementation of Hasselmann's paradigm for stochastic climate modelling based on stochastic Lie transport. Nonlinearity **36**(9), 4862 (2023)
17. Crisan, D., Holm, D., Lang, O., Mensah, P., Pan, W.: Theoretical Analysis and Numerical Approximation for the Stochastic thermal quasi-geostrophic model. Stochastics and Dynamics (2023)

© The Author(s), under exclusive license to Springer Nature Switzerland AG 2024
D. Goodair, D. Crisan, *Stochastic Calculus in Infinite Dimensions and SPDEs*,
SpringerBriefs in Mathematics, https://doi.org/10.1007/978-3-031-69586-5

18. Crisan, D., Lang, O.: Local well-posedness for the Great Lake Equation with transport noise. Revue Roumaine de Mathematiques Pures et Appliquees **66**(1), (2021)
19. Crisan, D., Lang, O.: Well-posedness properties for a stochastic rotating shallow water model. J. Dyn. Differ. Equ., 1–31 (2023)
20. Da Prato, G., Zabczyk, J.: Stochastic Equations in Infinite Dimensions, vol. 152. Cambridge University Press (2014)
21. Doob, J.L.: Measure Theory, vol. 143. Springer Science & Business Media (2012)
22. Drivas, T.D., Holm, D.D., Leahy, J.M.: Lagrangian averaged stochastic advection by Lie transport for fluids. J. Stat. Phys. **179**(5–6), 1304–1342 (2020)
23. Flandoli, F.: Topics on Regularization by Noise. Lecture Notes, University of Pisa (2013)
24. Flandoli, F.: An open problem in the theory of regularization by noise for nonlinear PDEs. In: Workshop Classic and Stochastic Geometric Mechanics, pp. 13–29. Springer (2015)
25. Flandoli, F., Galeati, L., Luo, D.: Delayed blow-up by transport noise. Commun. Partial Differ. Equ. **46**(9), 1757–1788 (2021)
26. Flandoli, F., Luongo, E.: Stochastic Partial Differential Equations in Fluid Mechanics, vol. 2328. Springer Nature (2023)
27. Flandoli, F., Mahalov, A.: Stochastic three-dimensional rotating Navier–Stokes equations: averaging, convergence and regularity. Arch. Ration. Mech. Anal. **205**(1), 195–237 (2012). http://doi.org/10.1007/s00205-012-0507-6
28. Flandoli, F., Pappalettera, U.: 2D Euler equations with Stratonovich transport noise as a large-scale stochastic model reduction. J. Nonlinear Sci. **31**(1), 1–38 (2021)
29. Flandoli, F., Pappalettera, U.: From additive to transport noise in 2D fluid dynamics. Stochastics Partial Differ. Equ. Anal. Comput., 1–41 (2022)
30. Galeati, L.: On the convergence of stochastic transport equations to a deterministic parabolic one. Stochastics Partial Differ. Equ. Anal. Comput. **8**(4), 833–868 (2020)
31. Galeati, L., Luo, D.: LDP and CLT for SPDEs with transport noise. Stochastics Partial Differ. Equ. Anal. Comput., 1–58 (2023)
32. Glatt-Holtz, N., Ziane, M., et al.: Strong pathwise solutions of the stochastic Navier-Stokes system. Adv. Differ. Equ. **14**(5/6), 567–600 (2009)
33. Glatt-Holtz, N.E., Vicol, V.C.: Local and global existence of smooth solutions for the stochastic Euler equations with multiplicative noise. Ann. Probab. **42**(1), 80–145 (2014). http://doi.org/10.1214/12-AOP773. https://doi-org.univaq.clas.cineca.it/10.1214/12-AOP773
34. Goodair, D.: Existence and Uniqueness of Maximal Solutions to a 3D Navier-Stokes Equation with Stochastic Lie Transport. Stochastic Transport in Upper Ocean Dynamics, p. 87 (2023)
35. Goodair, D.: Navier-Stokes equations with Navier boundary conditions and stochastic lie transport: well-posedness and inviscid limit. arXiv preprint arXiv:2308.04290 (2023)
36. Goodair, D.: Weak and strong solutions to nonlinear SPDEs with unbounded noise. arXiv preprint arXiv:2401.10076 (2024)
37. Goodair, D., Crisan, D.: On the 3D Navier-Stokes equations with stochastic lie transport. In: Stochastic Transport in Upper Ocean Dynamics II: STUOD 2022 Workshop, London, UK, September 26–29, vol. 11, p. 53. Springer Nature (2023)
38. Hairer, M.: An introduction to stochastic PDEs. arXiv preprint arXiv:0907.4178 (2009)
39. Hofmanová, M., Lange, T., Pappalettera, U.: Global existence and non-uniqueness of 3D Euler equations perturbed by transport noise. Probab. Theory Relat. Fields **188**(3), 1183–1255 (2024)
40. Holden, H., Karlsen, K.H., Pang, P.H.: The Hunter–Saxton equation with noise. J. Differ. Equ. **270**, 725–786 (2021)
41. Holden, H., Karlsen, K.H., Pang, P.H.: Global well-posedness of the viscous Camassa–Holm equation with gradient noise. arXiv preprint arXiv:2209.00803 (2022)
42. Holm, D.D.: Variational principles for stochastic fluid dynamics. Proc. R. Soc. A Math. Phys. Eng. Sci. **471**(2176), 20140,963 (2015)
43. Jakubowski, A.: On the Skorokhod topology. In: Annales de l'IHP Probabilités et statistiques, vol. 3, pp. 263–285 (1986)
44. Karatzas, I., Shreve, S.: Brownian Motion and Stochastic Calculus, vol. 113. Springer Science & Business Media (1991)

45. Kreyszig, E.: Introductory Functional Analysis with Applications, vol. 17. Wiley (1991)
46. Kunita, H., Watanabe, S.: On square integrable martingales. Nagoya Math. J. **30**, 209–245 (1967)
47. Lang, O., Crisan, D.: Well-posedness for a stochastic 2D Euler equation with transport noise. Stochastics Partial Differ. Equ. Anal. Comput. **11**(2), 433–480 (2023)
48. Lang, O., Crisan, D., Mémin, É.: Analytical properties for a stochastic rotating shallow water model under location uncertainty. J. Math. Fluid Mech. **25**(2), 29 (2023)
49. Lange, T.: Regularization by noise of an averaged version of the Navier–Stokes equations. J. Dyn. Differ. Equ., 1–26 (2023)
50. Bethencourt de Leon, A.: On the effect of stochastic Lie transport noise on fluid dynamic equations. PhD Thesis (2021)
51. Liu, W., Röckner, M.: Stochastic Partial Differential Equations: An Introduction. Springer (2015)
52. Lototsky, S.V., Rozovsky, B.L., et al.: Stochastic Partial Differential Equations. Springer (2017)
53. Luo, D., Saal, M.: A scaling limit for the stochastic mSQG equations with multiplicative transport noises. Stochastics Dyn. **20**(06), 2040,001 (2020)
54. Mikulevicius, R., Rozovskii, B.L.: Global L_2-solutions of stochastic Navier–Stokes equations. Ann. Probab. **33**(1), 137–176 (2005). http://doi.org/10.1214/009117904000000630
55. Oksendal, B.: Stochastic Differential Equations: An Introduction with Applications. Springer Science & Business Media (2013)
56. Pardoux, É., et al.: Stochastic Partial Differential Equations: An Introduction. Springer (2021)
57. Prévôt, C., Röckner, M.: A Concise Course on Stochastic Partial Differential Equations. Lecture Notes in Mathematics, vol. 1905. Springer, Berlin (2007)
58. Röckner, M., Schmuland, B., Zhang, X.: Yamada-Watanabe theorem for stochastic evolution equations in infinite dimensions. Condens. Matter Phys. **11**(2), 247 (2008)
59. Röckner, M., Shang, S., Zhang, T.: Well-posedness of stochastic partial differential equations with fully local monotone coefficients. arXiv preprint arXiv:2206.01107 (2022)
60. Rozovsky, B.L., Lototsky, S.V.: Stochastic Evolution Systems: Linear Theory and Applications to Non-linear Filtering, vol. 89. Springer (2018)
61. Tang, H.: On stochastic Euler-Poincaré equations driven by pseudo-differential/multiplicative noise. J. Funct. Anal., 110075 (2023)

Index